science

D1794944

GCSE Additional Applied Science

Course directors Jenifer Burden
Andrew Hunt
Robin Millar

Course editors Peter Campbell
Andrew Hunt

Module editor Caroline Shearer

Authors Ken Gadd
Anna Grayson
Andrew Hunt
Mike Tingle

4 Harnessing chemicals

OCR
RECOGNISING ACHIEVEMENT

Nuffield
Curriculum Centre

THE UNIVERSITY of York

OXFORD

Contents

Introduction

Chemistry is about making things. Chemists take raw materials from the world around us and turn them into products that help to meet our everyday needs. Chemicals help to keep food fresh; they prevent and cure disease; and they brighten and protect our homes.

In this module you will learn about the techniques for making new chemicals and then mixing them to make useful products. You will also learn about the people in a range of organisations that help us all by harnessing chemicals.

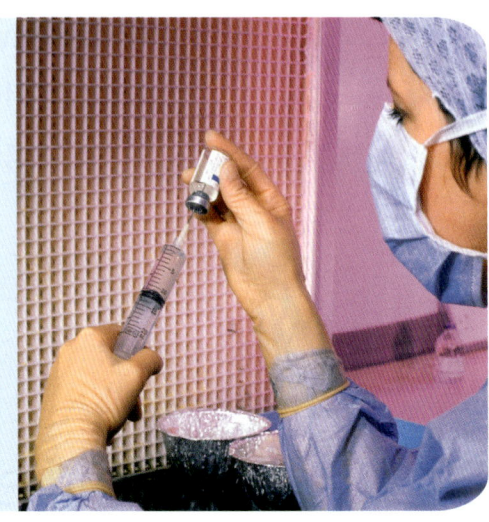

Creating harmony...

Perfume chemicals

André sits at the organ composing … a perfume! He's an artist as much as a scientist. His work is precise, but his main 'tool' is his sensitive nose. He can recognise 3000 different aromas. The organ is what perfumers call their rows of bottles containing essential oils.

It may take months to get the composition of a fragrance just right, to create perfect harmony between the various parts.

The right chemistry

Creating a perfume is an art, but it relies on science. André's aroma organ depends on chemical methods to extract essential oils from plants. And when André creates a new formulation for a perfume, it is scientists who reproduce his mixture exactly, on an industrial scale.

Chemists use solvents to extract essential oils from flowers, leaves, and fruit. The oil dissolves in the solvent to give a dilute solution. Distilling off the solvent concentrates the oil. The solvent can be re-used. The choice of solvent and temperature are controlled to avoid damaging the scent of the oil.

To create a new perfume, André mixes various oils, gradually altering the mixture's composition. He dips paper strips into each mixture to compare the smells. He records the exact mass of each oil, so when he produces an aroma he likes, he knows how he made it.

Stonewashed jeans

Comfortable, hard-wearing, and good-looking – denim jeans have long been the favourites of many people, starting with the American cowboys.

Jeans – fashion and function

Jeans were popular with cowboys because they were very hard-wearing. Traditionally the dye for colouring jeans has been indigo.

Indigo has been used for centuries. It was first found in plants like woad. It could be extracted from the plants and isolated as a blue solid. Now synthetic indigo has almost entirely replaced indigo from natural sources. Throughout the world the chemical industry manufactures around 17 000 tonnes of indigo each year, almost all of which is used on denim. Forty per cent of the dye is produced by the German company BASF.

Naphthalene to Indigo

a naphthalene molecule

a molecule of indigo

Chemists have devised clever ways to modify molecules. The synthesis of indigo from naphthalene takes five steps.

Seeds of
dyer's woad

Colour from crude oil

The starting chemical for making indigo is naphthalene. This is an organic compound which is obtained from coal or crude oil.

Indigo is insoluble in water. So before it can be used to dye clothes it must be changed into a form that dissolves in water. Chemists use a chemical called sodium dithionite to convert indigo to a water-soluble form. This is almost colourless, and can soak into the denim of jeans.

The indigo turns back into its insoluble blue form when exposed to oxygen. This oxidation reaction produces fine specks of the dye trapped in the fibres of the jeans so that they cannot easily wash out.

Brute force or enzymes?

Some people prefer jeans with a 'stonewashed' appearance. The dyed denim is put in a rotating drum, rather like a washing machine, along with some stones. The stones break some of the fibres and release the indigo dye. The result is jeans that look faded and worn.

This is quite a brutal process. Tough machines are needed! Scientists have come up with a gentler alternative. An enzyme called cellulase can digest cotton fibres. The enzyme is a natural catalyst. It speeds up a reaction which breaks apart the cellulose polymer in cotton. Indigo is released as the surface fibres break down.

Plants such as woad were the source of indigo until chemists found a way to synthesise the dye in 1897.

5

sweeter than sugar

Sweeteners old and new

The sugar that we add to food and drink is extracted from sugar cane or sugar beet. Eating too much sugar is not good for you, yet most people find it hard to resist the sweet taste.

Instead of sugar, a substance that is 300 times sweeter has been used to sweeten food and drink for over 100 years. This substance is saccharin, the first artificial sweetener to be discovered. It has the formula $C_7H_5NO_3S$. Saccharin is known as a calorie-free sweetener because the digestive system does not break it down to provide energy.

A chance discovery

Saccharin can be made from toluene by a series of complicated, but fairly cheap, steps. Toluene is an organic compound obtained from coal or crude oil. The taste of saccharin was discovered when a chemist did not wash his hands on leaving the laboratory (poor laboratory practice!).

He tasted the intense sweetness later when eating some bread. He realised that he was working on something very interesting and went on to find out exactly what chemical it was.

Saccharin is 300 times as sweet as sugar.

Designer colours

In the 1920s, Ford painted their cars 'any colour you like, so long as it's black'. Nowadays our world is much more colourful, thanks to chemistry – and physics, because colour depends on chemicals that absorb and reflect different colours of light. We really can make things any colour we like; and any shade of that colour.

The pigments in paints are insoluble chemicals. Paint-makers mix them with liquids to make thick suspensions which spread smoothly over surfaces. Pigments mixed with polymers make coloured plastics.

It takes more than just a good pigment to make a paint. The other ingredients allow the colour to be applied successfully, and then make sure that it dries to an even film, sticks to the surface, and stands up to wear and tear.

It is not only the colour which matters. The paint must spread evenly, dry fast, and form a durable surface. The technician in protective clothing is putting the finishing touches to the paintwork on a new truck.

A paint contains four main ingredients:
- pigment - the colouring material
- binder - holds the pigment particles together and forms a tough layer when paint 'dries'
- solvent - dissolves the binder so the paint can be spread; may be water or an organic solvent
- additives - such as fungicides to stop mould growing

Producing the right mixture is another example of the art of formulation, like mixing perfumes.

What's in a toothpaste?

The formula for success

Oxen hooves, heated eggshells, pumice, and water – this is the Ancient Egyptians' formula for toothpaste. Grinding the ingredients together produced an abrasive to clean teeth. It worked like sandpaper.

The first 'tooth-cleaning powder' was used in the UK in the eighteenth century. People mixed it with water and rubbed it on their teeth. These early pastes were very abrasive and could damage teeth.

Modern toothpastes are gentler on the teeth. Their formulation ensures they are better at preventing tooth decay. Manufacturers continue to experiment to find better ways of protecting teeth.

What's in your toothpaste?

Check the ingredients in your favourite toothpaste. There may be a long list but they usually include:

- water along with chemicals to keep the toothpaste moist (humectants) – these make up 75% of the toothpaste

- an abrasive such as silica or calcium carbonate to remove plaque – about 20%

- colouring, flavouring, foaming agents, a gum to hold the ingredients together, and a preservative – about 3%

- a source of fluoride such as sodium fluoride or sodium monofluorophosphate – about 0.2% fluoride

Toothpastes for people with sensitive teeth also contain strontium chloride or potassium nitrate. After a few weeks of regular use, these chemicals make the nerve endings in the gums less sensitive.

The BDA

Dentists belong to the British Dental Association (BDA). The BDA gives advice to the public. They also have a panel of experts who test toothpastes: they check that pure and safe chemicals have been mixed in the correct proportions. If the toothpaste is approved the manufacturer can put the BDA logo on the box.

The association's experts are convinced by the evidence that toothpastes containing sodium fluoride, sodium monofluorophosphate, or a combination of the two are effective in preventing tooth decay.

Researchers at Portsmouth University have suggested that toothpaste made from crabs' shells could reduce dental infections. Chemicals in the shells stick to teeth and gums. Mixed with water they form triclosan, a chemical that can kill bacteria. The challenge is to make particles of the right size and formulate them with other substances to make the toothpaste. However, the BDA thinks that more needs to be known about the paste.

Do you follow the guidelines?

The BDA says . . .

▶ Brush your teeth twice a day using fluoride toothpaste. Fluoride prevents tooth decay.

▶ Rinse briefly with a small amount of water.

▶ Choose toothpastes that contain triclosan with either copolymer or zinc citrate. These help control plaque and keep gums healthy.

People and organisations

The chemical industry converts raw materials into pure chemicals. Its raw materials include crude oil, natural gas, minerals, air, and water.
The chemicals it produces are used to make valuable products. In this section you will meet some of the companies processing chemicals, and some of the people that work in them. Many different skills are required to harness chemicals.

Chemicals from rocks and minerals

Huge multinational companies mine and quarry millions of tonnes of **minerals** every year. The minerals they extract include metal **ores**, limestone, salt, and sulfur. These companies employ people with a range of scientific skills in three main areas: prospecting, mining, and processing.

Geologists use their knowledge of rocks and sensitive instruments to find minerals buried beneath the ground.

A few chemicals such as limestone, salt, sulfur, and gold come straight from the ground. They just need purifying. In most cases, however, mining is just the starting point. Scientists use chemical processing to convert minerals into useful **inorganic chemicals**.

The chemical industry heats limestone in furnaces to make quicklime. Heating converts the calcium carbonate to calcium oxide. Calcium oxide is used by farms to improve soil fertility, by power stations to remove pollutants, and by steelmakers to remove impurities.

The chemical industry converts sulfur into sulfuric acid. Sulfuric acid is widely used as a general purpose acid. It is involved in the making of paints, pigments, fertilisers, detergents, and plastics.

The chemical industry converts sodium chloride (salt) solution into chlorine, hydrogen, and sodium hydroxide. Chlorine is used to make many chemicals including bleach and the polymer PVC. Sodium hydroxide is the most widely used alkali.

The chemical industry converts phosphate rock into phosphoric acid. Phosphoric acid is used to make fertilisers, washing powders, food additives, and ingredients for toothpaste.

Question

1 Draw up a table to show examples of the minerals used to make chemicals. Use these headings:

Mineral; Chemicals made; Uses

Key words

minerals
ores
inorganic chemicals

Harvesting chemicals

Not all chemicals are extracted from minerals by large industrial companies. Norfolk Lavender is a family firm which produces perfume chemicals from lavender. Linn Chilvers, the son of a nursery gardener, started the business in 1932. He bought copper stills from France, the traditional source of the perfume, so he could extract perfume oils from his Norfolk lavender crop.

Harvesting lavender in Norfolk

The perfume oils are examples of **organic chemicals**. They are carbon compounds. Lavender extracts are used in perfumes, soaps, talcs, and many other products.

A chemist from Leicester, Mr Avery, saw the publicity about the new business and contacted Linn. He had a recipe for a lavender perfume made for King George IV. For many years Mr Avery travelled to Norfolk to mix the essences for the Norfolk Lavender perfume.

The business grew. By the early 1990s they were selling 60 products to more than 25 countries. In 1977 there were about 12 staff: now there is a team of 120 full-time staff.

Key words

organic chemicals
batch process

Heating lavender with steam distils off the perfume chemicals. This is a **batch process**. Separate batches of lavender are loaded into the copper containers and then distilled.

The chemical industry

What is the chemical industry?

The chemical industry converts **raw materials** into useful products. The products include chemicals for use as drugs, fertilisers, detergents, paints, and dyes.

The chemical industry makes **bulk chemicals** on a scale of thousands or even millions of tonnes per year. Examples include ammonia, sulfuric acid, sodium hydroxide, chlorine, and ethene.

On a much smaller scale the industry makes **fine chemicals** such as drugs and pesticides. It also makes small quantities of **speciality chemicals** needed by other manufacturers for particular purposes. These include such things as flame retardants, food additives, and liquid crystals for flat-screen televisions.

The part of a chemical works which produces a chemical is called a 'plant'. Some of the chemical reactions are carried out at a high temperature, so a source of energy is needed to heat them. Electrical power is also needed for pumps to move reactants and products from one part of the plant to another. Sensors monitor the conditions at key points in the plant. The data from these is fed to computers in the control centre where the technical team controls the plant.

Key words

raw materials
bulk chemicals
fine chemicals
speciality chemicals
scale up

The chemical industry converts raw materials into pure chemicals such as acids, alkalis, salts, solvents, compressed gases, and organic compounds.

dyes and pigments 3%
industrial glass 5%
3% agrochemicals
basic inorganics 2.5%
8% paints, varnishes, and printing inks
basic organics 12%
31.5% pharmaceuticals
fertilisers 1%
plastics and synthetic rubber 7.5%
11.5% soaps, toiletries, and cleaning preparations
synthetic fibres 2%
13% other specialities

The pie chart shows the range of products made by the chemical industry in Britain, and their relative values.

Questions

1 Give the name and chemical formula (see pages 58–9) of a bulk chemical.

2 What percentage value of products of the chemical industry in Britain are used

   a in agriculture and horticulture
   b to make polymers
   c for medical diagnosis and treatment?

3 List these chemicals under two headings: 'bulk chemical' and 'fine chemical'.

   ▸ the drug aspirin
   ▸ the hydrocarbon ethene
   ▸ the perfume chemical citral
   ▸ the acid sulfuric acid
   ▸ the herbicide glyphosate
   ▸ the alkali sodium hydroxide
   ▸ the food dye carotene
   ▸ the pigment titanium dioxide

People in the chemical industry

People with many different skills are needed in the industry. Research chemists work in laboratories to find new processes and make new products. They work closely with people in the marketing and sales department who find out whether people want the product. If the new product is promising it may first be tried out by making it in a pilot plant before full-scale production starts.

As part of the market research, possible new products are given to customers for trial. At the same time financial experts estimate the value of the new product in the market and compare this with the cost of making it, to check that the new process will be profitable.

Chemical engineers have to **scale up** the process and design a full-scale plant. This can cost hundreds of millions of pounds.

Some products from the chemical industry go directly on sale to the public, but most of them are used to make other products. Transport workers carry the chemical products to the industry's customers.

Every chemical plant needs managers and administrators to control the whole operation. There are also people in service departments who look after the needs of the people working there, such as medical and catering staff, and training and safety officers.

Questions

4 Draw up a table to summarise who works in the chemical industry and what they do. Use these headings:

Type of worker
What they do

5 When the tank in the picture below is in use, it is filled with a reaction mixture. Suggest the purpose of

a the rotating paddle in the centre of the tank

b the network of pipes round the edge of the tank

Plant operators monitor the processes from the control room.

Maintenance workers help to keep the plant running.

The largest refinery in the UK

Arriving at an industrial site can be a shock. These young engineers have started work at the Fawley oil **refinery**.

I had no idea of the sheer size of an oil refinery. Now I'm one of the team running it. What a responsibility!

The pipework and towers are huge. I am beginning to understand what is meant by scaling up.

Many people share my worries about pollution. It's great to feel that my work will be about keeping up environmental standards. After all, we will rely on oil for many more years.

Crude oil is a mixture of organic chemicals called **hydrocarbons**. The oil refinery separates and purifies the mixture by distilling it. The products are fuels, lubricants, and raw materials for the chemical industry.

Fawley refinery near Southampton is the largest in the UK. Its terminal on Southampton Water sees around 2000 ships come and go every year, handling 22 million tonnes of crude oil and other products. As well as all the pipes and towers there are workshops, laboratories, computer centres, administration blocks, and a health centre. Most of the **processes** carried out there are **continuous**. They are kept running day and night, seven days a week, with the crude oil flowing in and the products flowing out continuously.

> **Key words**
>
> refinery
> hydrocarbons
> continuous process

> **Questions**
>
> 1 Is crude oil a renewable or non-renewable resource?
>
> 2 Find three things that impressed the young engineers when they came to the Fawley refinery.

The Fawley refinery in Southampton

Sustainability

The chemical industry aims to become more **sustainable** by

- developing **renewable resources**
- making efficient use of energy
- reducing waste

Energy saving

Many chemical reactions give out energy. They are **exothermic**. Burning is an obvious example. Chemical engineers aim to capture the energy from exothermic reactions and convert it to steam or electricity. Sometimes this provides all the energy needs of the chemical plant.

Waste is not wasted

All chemical processes give a mixture of products. One the chemists want to make, and others are called by-products. The process is more sustainable if the by-products can be recycled or otherwise used.

Huntsman Tioxide, a company based in Grimsby, have found ways of cutting down waste. The company makes titanium dioxide, the world's most important white **pigment**. The pigment is used in paint, to coat paper, as a filler in plastics, and in cosmetics and toothpaste.

Huntsman Tioxide makes titanium dioxide from ilmenite using sulfuric acid. Ilmenite is an **oxide** of iron and titanium, with the formula $FeTiO_3$. The main by-product is iron sulfate. One use for iron sulfate is to treat drinking water. Neutralising an acidic solution of iron sulfate with calcium oxide produces 'red gypsum' which can be used as a soil conditioner on local farms.

There is a large volume of sulfuric acid left at the end of this process. Neutralising the acid with calcium carbonate makes calcium sulfate, or 'white gypsum'. The company used to dump most of the white gypsum, but now it is used to make plaster. So less waste is sent to landfill sites, and less natural gypsum needs to be mined.

Key words

sustainable
renewable resources
exothermic
pigment
oxide

Questions

1 State three ways of making a chemical process more sustainable.

2 Which of these ways of making processes more sustainable have been applied by Huntsman Tioxide?

3 Draw a flow diagram to illustrate titanium dioxide production, showing how the by-products are used.

Titanium dioxide is the white pigment in this paint protecting a railway bridge in Newcastle-upon-Tyne.

Huntsman Tioxide's plant at Grimsby. The company is the world's third largest producer of titanium dioxide.

Sustainable pain relief

Over-the-counter drugs

Aspirin, paracetamol, ibuprofen, and codeine are four pain-relieving **drugs**. They are safe enough to be sold 'over the counter' without a prescription. These drugs help to give relief from headache, toothache, and period pain.

Over-the-counter pain relievers may be sold under their own names or as branded **medicines**, such as Anadin, Nurofen, and Panadol. These products are special **formulations**. They may contain more than one active ingredient, together with other ingredients such as binders to hold the tablets together, or salts to make tablets fizz when put into water.

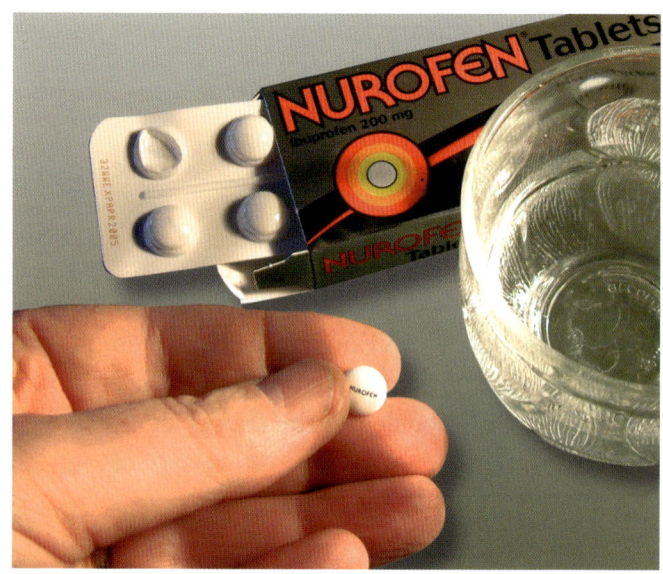

A better synthesis

Boots patented ibuprofen in the 1960s. At that time there were six stages in the complex processes used to make the drug. It was a very wasteful process with a lot of unwanted by-products.

The patent ran out after 20 years, when other companies were allowed to make and sell the painkiller. A more efficient **synthetic route** was developed. The new process has only three stages, needs fewer chemicals, and produces fewer harmful by-products. It is a 'greener', more sustainable, way of making ibuprofen.

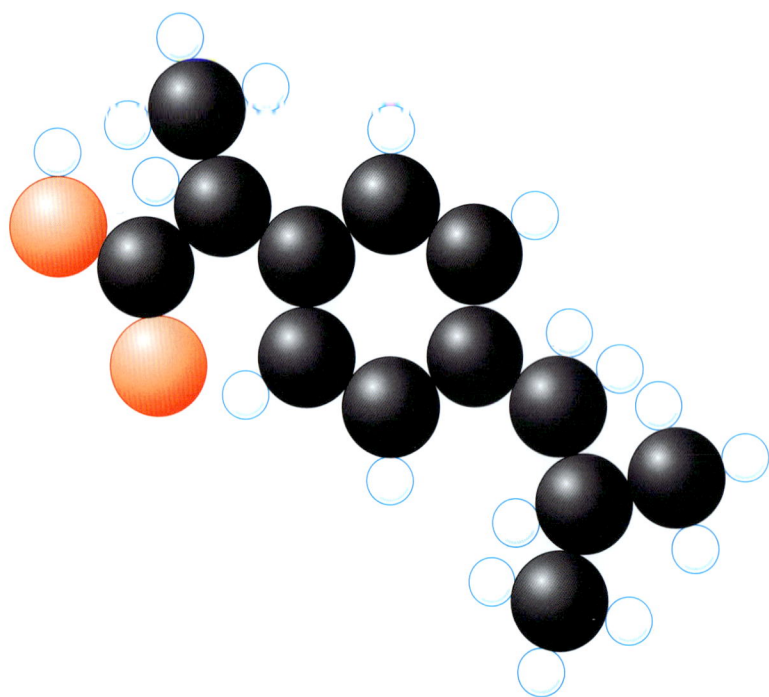

A model of an ibuprofen molecule. Black = carbon, white = hydrogen, red = oxygen.

Questions

1 Analgesics are painkillers. Name two analgesic chemicals.

2 Explain the difference between an 'over-the-counter' and a 'prescription' medicine.

3 Give a reason why a medicine may contain more than one ingredient.

4 In a molecule of ibuprofen how many atoms are there of each type?

  a carbon

  b oxygen

  c hydrogen

Key words

drug
medicine
formulation
synthetic route

Developing a new medicine

Laboratory research and testing

Most medicines are very expensive because developing new drugs is a long and costly process. It can take over 15 years and £300 million to develop a new drug. Thousands of new compounds are invented, made, and tested. Only a few are chosen for trials with patients. Eventually one may be manufactured and used in a medicine.

Clinical research

The compounds chosen for trials are given to patients to find out how effectively each drug treats a particular disease. This is clinical testing. Researchers also check for harmful side-effects. On the basis of the trials, the most effective drug with few or no side-effects is chosen.

Process research

The company has to find a way to manufacture a new product economically and safely. The methods used in the research laboratory are not usually suitable for large-scale production. Industrial chemists find an economic, safe, and sustainable synthetic route for scaling up the production.

Formulation

Each dose of a medicine may only contain a few milligrams of the active ingredient. The job of the pharmacologists is to discover the best way to get the drug to the right part of the body. This might be as a liquid for injection, or a cream to be absorbed through the skin, or a coated tablet to slide down the throat. The challenge is to find a formulation that is effective, acceptable to patients, and also keeps well.

This pilot plant uses a batch process to make a few kilograms of a drug at a time. The product is used for further clinical trials.

Questions

1 Explain why new medicines are expensive.

2 What is the difference between '*in vitro*' and '*in vivo*' screening of drugs?

3 Suggest a reason why companies give their medicines brand names.

Manufacture and quality assurance

Eventually the medicine goes into production. Plant managers and operators ensure the smooth running of the process. Analytical scientists check that both the chemicals used as reactants, and the final products, meet the necessary standards.

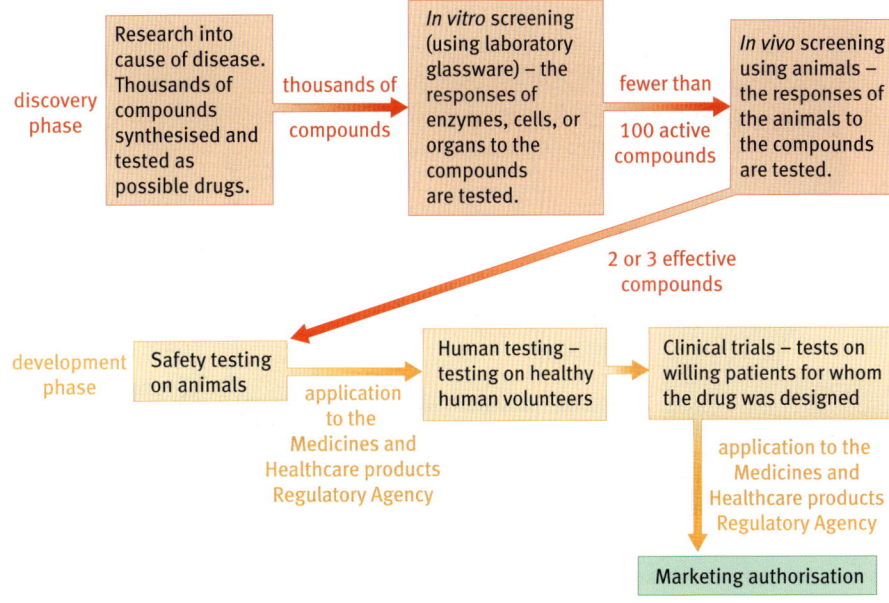

discovery phase — Research into cause of disease. Thousands of compounds synthesised and tested as possible drugs. → thousands of compounds → *In vitro* screening (using laboratory glassware) – the responses of enzymes, cells, or organs to the compounds are tested. → fewer than 100 active compounds → *In vivo* screening using animals – the responses of the animals to the compounds are tested.

2 or 3 effective compounds

development phase — Safety testing on animals → application to the Medicines and Healthcare products Regulatory Agency → Human testing – testing on healthy human volunteers → Clinical trials – tests on willing patients for whom the drug was designed → application to the Medicines and Healthcare products Regulatory Agency → Marketing authorisation

Researching, developing, and testing a new drug is a long and expensive process.

Safe practice

There are many European and national regulations designed to protect the general public, people at work, and the environment. The chemical industry has to make sure that it complies with these regulations.

In the UK, the Department for Trade and Industry (DTI) provides guidance for chemical companies on the relevant regulations. Their advice covers seven main areas of regulation:

- ▶ chemical products – the classification of chemicals, and their packaging and labelling

- ▶ transport – the transport of chemicals by road, rail, air, or sea

- ▶ worker safety – health and safety at work, including the control of substances hazardous to health

- ▶ building and engineering – the relationship of the design of buildings to the chemicals in them

- ▶ storage – the safe storage of chemicals in bulk

- ▶ pollution control – environmental protection

- ▶ waste disposal – the assessment and disposal of waste

If a tanker is carrying a hazardous substance such as petrol, it should carry an orange sign such as this.

Non-hazardous cargo still needs to be labelled, in case of an accident.

Transport workers bring materials in and out of a chemical plant. The Hazchem labelling code is compulsory for tankers so that the emergency services know about possible hazards if the load is spilled.

Questions

1 Where could a chemical company director find advice on the regulations about the storage of chemicals?

2 Why do tankers carrying chemicals have to be labelled with the Hazchem code?

3 What containers, other than road tankers, carry hazard warning signs?

Worker safety

The Health and Safety Executive (HSE) aims to protect people's health and safety by making sure risks in the workplace are properly controlled. Its staff come from a range of different backgrounds, including administrators, lawyers, statisticians, inspectors, scientists, technicians, engineers, doctors, and nurses.

The HSE works with large companies and small businesses in three main areas:

- accident investigation – they collect information about workplace accidents, to help draw up new advice on how to reduce risk

- guidance – they advise companies on health and safety issues

- inspection – they check safety standards and make sure companies obey the law

The regulations

The HSE enforce many regulations designed to protect people at work. Here are two examples of recent regulations.

Control of Substances Hazardous to Health (COSHH)

Employers must do a risk assessment for chemicals or mixtures that could be hazardous to health, and make sure that proper controls and precautions are followed. This could mean changing procedures to use less hazardous substances.

Personal Protective Equipment at Work

If hazards cannot be removed completely the employer must provide suitable personal protective equipment (PPE) and give training in its use. PPE includes protective clothes, boots, gloves, safety helmets, and goggles.

Questions

1 What organisation ensures that companies follow safety regulations?

2 Suggest a reason why the workers in the photograph have not tucked their trousers into their boots.

3 What PPE is available in your school science laboratory?

These workers in a factory are wearing overalls and rubber gloves to protect their skin from chemicals. Their masks filter the air they breathe, and also protect their eyes.

Making and testing formulations

Preparing, separating, and purifying chemicals is only the first step in making a usable product. Most everyday products are complex mixtures, made to precise formulations.

Trade secrets

Danisco makes ingredients for the food and drink industry. Among its products are concentrates which can be diluted to produce carbonated drinks. Food companies buy concentrates, dilute them with carbonated water, then pasteurise and bottle the drinks. After stamping with a sell-by date the bottles are sent to the shops.

Barry Taylor has been in the food industry all his working life. He specialises in flavourings and soft drinks for Danisco. Most of Barry's formulations are secret, but the diagram on the next page shows how a typical pinapple and grapefruit fizzy drink is made.

Barry Taylor in his laboratory

Testing formulations

Danisco perform tests on their formulations, to make sure that they meet their own quality control standards and all the required national and international safety standards.

Food standards

Manufactured products must pass strict quality assurance tests to make sure they are 'fit for purpose' (do what they are designed to do), and are safe if used correctly. For most products there are national and international standards. The Food Standards Agency (FSA) sets safety standards for food products in the UK, including the use of chemical additives. The type and amount of additive used is regulated together with the way it is listed on the food label.

Local authorities sometimes carry out tests to check that food on sale in the shops is safe. These tests are performed by public analysts. Tests follow **standard procedures**, so that they are the same wherever the tests are done.

Questions

1 Make a table to show the ingredients of ten litres of fruit concentrate. Use these headings:

Ingredient   Quantity
Reason for adding the ingredient

2 Draw a flow diagram to show the stages from the starting ingredients to the final product – bottles of fizzy drink in a refrigerator at home.

Formulating a concentrate for a fruit drink

Sweeteners

Aspartame (137.0 g) Aspartame is 200 times sweeter than sugar. It is most stable in slightly acidic solutions (pH 4.3).

Sodium saccharin (90.7 g) Saccharin is 300 times sweeter than sugar, but has a distinctive after-taste which not everyone likes. Saccharin itself is not all that soluble, so the more soluble compound sodium saccharin is used.

Artificial sweeteners are used instead of sugar as this is a low-calorie drink.

Colouring

Lutein (50 g) Colouring is added to 'maximise the enjoyment of the beverage', or in other words to make the drink look good. Only compounds that have been proved to be safe and non-toxic can be used. Barry recommends lutien, a natural extract of Aztec marigold flower, to get the right sort of yellow colour.

Preservative

Sodium benzoate, 20% solution (206.0 ml) This is a preservative to stop the growth of moulds and bacteria. It would not be necessary if the drink was for immediate consumption. Benzoic acid is found naturally in some fruit (such as cranberries), but it is not very soluble in water. The salt sodium benzoate is more soluble and also acts as a preservative.

Fruit and flavouring

Pineapple and grapefruit compound (12.5 litres). This is a complicated mixture of pineapple and grapefruit juice, fruit flavouring, citric acid (found naturally in grapefruit), and ascorbic acid (vitamin C). The acids add flavour and also make the drink more refreshing. The fruit flavouring is added to intensify the taste. The peel of citrus fruit contains highly flavoured oils that do not mix with water, so citrus flavouring is made by mixing the oils with an **emulsifying agent** which disperses the oil in the mixture.

Water

Purified water (to make the volume up to 100 litres) Barry recommends soft water, so tap water is treated to remove any dissolved minerals. It is then passed through a series of filters and treated with UV radiation to remove suspended organic material and kill any remaining micro-organisms.

Key words
standard procedure
emulsifying agent

The science

A chemical language

Symbols

Chemists use symbols for the chemical **elements**. Page 58 shows some examples. One advantage of symbols is that they are the same all over the world.

Formulae

Every compound has a **chemical formula**. The formula shows the elements that make up the compound. The formula also shows the number of atoms of each type in the formula. For example:

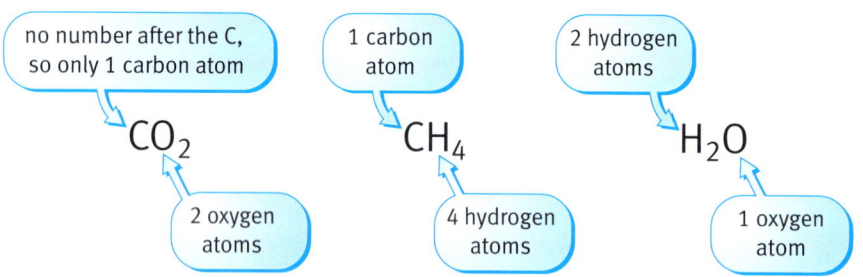

no number after the C, so only 1 carbon atom

CO_2

2 oxygen atoms

1 carbon atom

CH_4

4 hydrogen atoms

2 hydrogen atoms

H_2O

1 oxygen atom

Chemists use symbols and formulae to describe chemical changes. They write two kinds of equation:

- word equations
- balanced symbol equations

Word equations

In a word equation the rule is to write the names of the reactants on the left and the products on the right. An arrow ($\rightarrow$) between them is short for 'goes to', meaning 'change into' or 'become'.

Example 1: The word equation for methane burning

Methane burns by reacting with oxygen in air.
The equation may be written:

methane + oxygen → carbon dioxide + water

Cooks burn methane in air to heat food. The methane reacts with the oxygen in the air. The products are water and carbon dioxide.

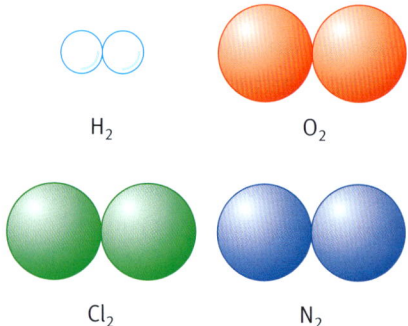

H_2 O_2

Cl_2 N_2

Some non-metals are made up of small molecules. These elements are usually represented by the symbols for these molecules. In any element all the atoms are the same. In these four elements the atoms are joined up in pairs.

Questions

1 Give the chemical names of these molecules:

 a H_2

 b O_2

 c Cl_2

 d N_2

2 Choose the correct chemical formula from the list for each answer.

 Na_2CO_3

 C_2H_4O

 Cl_2

 $C_2H_4O_2$

 Which chemical

 a is an element

 b contains sodium

 c contains carbon, oxygen, and hydrogen in the ratio 2:1:4?

Balanced symbol equations

Writing an equation in symbols makes it much easier to see what is happening chemically during a reaction. Atoms are not created or destroyed during reactions. They are simply rearranged. This means that the number of atoms of each type must be the same on both sides of the equation. In this sense the two sides of the equation are 'equal' and that is why a chemical equation must always 'balance'.

Rules for writing balanced equations

STEP 1 Write down a word equation.

STEP 2 Underneath write down the correct formula for each reactant and product.

STEP 3 Check to see if the equation needs balancing.

STEP 4 Balance the equation if necessary, by putting numbers in front of the formulae.

STEP 5 Add state symbols.

NEVER change the formula of a compound or element to balance the equation.

Follow these rules for balancing equations.

Example 2: A balanced symbol equation for the reaction of a solution of sodium hydroxide with hydrochloric acid

Step 1

sodium hydroxide + hydrochloric acid → sodium chloride + water

Step 2

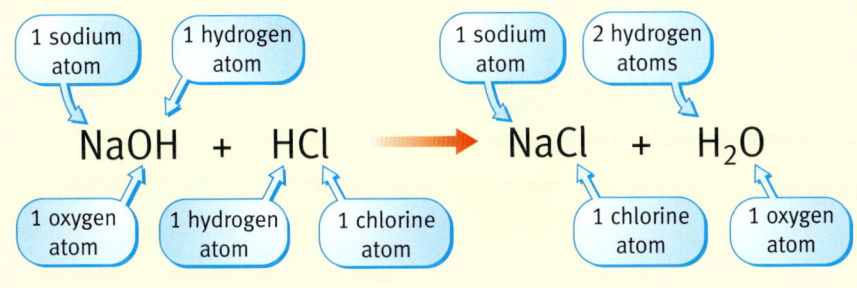

1 sodium atom | 1 hydrogen atom | 1 sodium atom | 2 hydrogen atoms

$NaOH + HCl \rightarrow NaCl + H_2O$

1 oxygen atom | 1 hydrogen atom | 1 chlorine atom | 1 chlorine atom | 1 oxygen atom

Step 3

1 sodium atom	1 sodium atom
1 oxygen atom	1 oxygen atom
(1 + 1) hydrogen atoms	2 hydrogen atoms
1 chlorine atom	1 chlorine atom

Step 4

$NaOH + HCl \rightarrow NaCl + H_2O$

The equation balances.

Step 5

$NaOH(aq) + HCl(aq) \rightarrow NaCl(aq) + H_2O(l)$

Question

1 Use the five steps to write balanced symbol equations for the following reactions. See pages 58–9 for the chemical formulae of some acids, antacids, and salts.

a potassium hydroxide solution with dilute nitric acid (the salt formed is KNO_3)

b magnesium metal with dilute hydrochloric acid (the salt formed is $MgCl_2$)

c solid zinc carbonate with dilute sulfuric acid (the salt formed is $ZnSO_4$)

d calcium hydroxide with dilute hydrochloric acid (the salt formed is $CaCl_2$)

Key words

elements
chemical formula
word equations
balanced symbol equations

Acids

Types of acid

Pure **acids** may be solids, liquids, or gases.

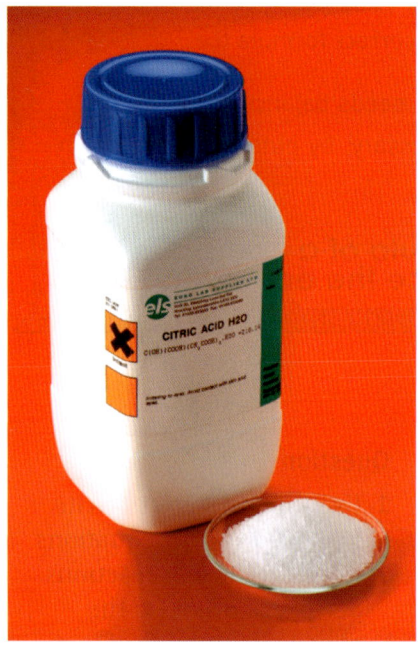

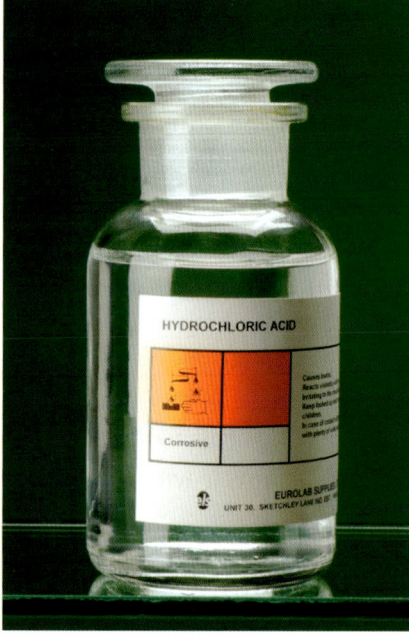

Citric acid is an example of an acid which is solid when pure. This is the acid which gives orange and lemon juices their sharp flavour.

Acetic acid is the acid in vinegar. Pure acetic acid is a liquid, though in an unheated store it can crystallise on cold winter days. Chemists call this compound ethanoic acid. Like all 'eth' compounds it contains just two carbon atoms in each molecule. It is related to the alcohol ethanol, and to the hydrocarbon ethane.

Hydrogen chloride is a gas. It dissolves in water to form hydrochloric acid. All common acids dissolve in water.

Inorganic acids

Three of the common laboratory acids are inorganic. They are sulfuric, hydrochloric, and nitric acids. The element sulfur is the starting point for making sulfuric acid. The chlorine in hydrochloric acid comes originally from common salt, which chemists call sodium chloride. The raw materials for making nitric acid are nitrogen and oxygen from the air together with water. Making nitric acid is a complex industrial process with several stages.

Organic acids

Acids are vital to life and there are many different acids in living things. Citric acid and acetic acid are just two of them. Other examples include palmitic and stearic acids which are used to make soap. Lactic acid is the chemical which makes muscles ache after long strenuous exercise. It is also the acid in sour milk.

Key words

acids
indicator
concentration
pH scale
salt
oxides

Acids and indicators

The easiest way to recognise an acid is to test a solution with an **indicator**. Litmus turns red in any acid. Full range or universal indicator turns to shades of orange and red in acid depending on the type of acid and the **concentration** of the solution.

The **pH scale** is a number scale which shows the acidity or alkalinity of a solution in water. Most laboratory solutions have a pH in the range 1–14. Solution of acids have a pH lower than 7.

Acids and metals

Metals such as magnesium, zinc, and iron react with dilute solutions of acids. The mixture of metal and acid fizzes as the reaction produces hydrogen gas. This is shown by the word equation:

metal + acid → **salt** + hydrogen

Some metals are unreactive with dilute acids. Examples are copper and lead.

Acids and carbonates

Another reaction which produces a gas is the reaction of a carbonate with an acid. Limestone is an example of a carbonate. Limestone and marble are both forms of calcium carbonate. These minerals fizz when mixed with dilute hydrochloric acid.

metal carbonate + acid → salt + carbon dioxide + water

Acids and metal oxides and hydroxides

Most metal **oxides** and metal hydroxides dissolve when they react with an acid to make a salt. No gas forms when metal oxides or metal hydroxides neutralise an acid.

metal oxide (or hydroxide) + acid → salt + water

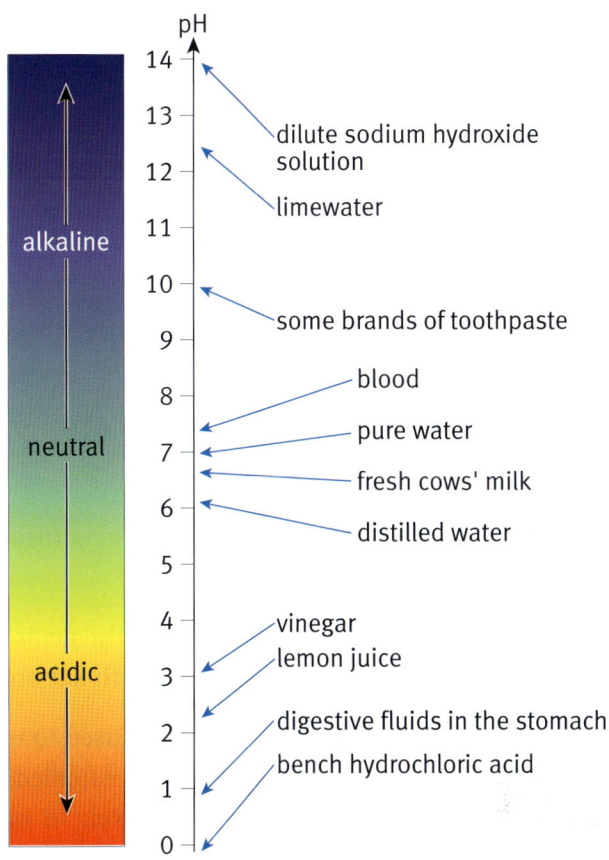

pH

- 14 — dilute sodium hydroxide solution
- 13
- 12 — limewater
- 11 — alkaline
- 10
- 9 — some brands of toothpaste
- 8 — blood
- 7 — neutral — pure water / fresh cows' milk
- 6 — distilled water
- 5
- 4 — vinegar
- 3 — acidic — lemon juice
- 2 — digestive fluids in the stomach
- 1 — bench hydrochloric acid
- 0

The pH scale

Geologists use hydrochloric acid to test rocks. If the rock fizzes it shows that the rock is a carbonate such as limestone or dolomite.

Questions

1 Write a word equation for the reaction of dilute sulfuric acid with

  a zinc

  b magnesium carbonate

  c sodium hydroxide

2 Write balanced symbol equations for the reactions in question 1 (see pages 58–9 for the formulae).

Alkalis

Pharmacists sell **antacids** to control heartburn and indigestion. The chemicals in these medicines are the chemical opposites of acids. They are designed to neutralise excess hydrochloric acid produced in the stomach – hence the name 'antacids'.

Some chemical antacids are soluble in water to give solutions with a pH above 7. Chemists call them **alkalis**. Common alkalis are sodium hydroxide and potassium hydroxide. These alkalis are not safe to swallow.

Types of alkali

Alkalis are used at home for two purposes: to neutralise acids and to remove greasy dirt.

Toothpaste is mildly alkaline. It neutralises the acids which attack teeth. These acids form when bacteria in the mouth act on sugars in food.

Milk of Magnesia is one antacid ingredient used to treat too much acidity in the stomach. Milk of Magnesia is supplied either as tablets or as a milky suspension of magnesium hydroxide in water.

Powerful household cleaners contain sodium or potassium hydroxide. These alkalis break down grease and can remove tough deposits in ovens. They have to be used with great care.

Alkalis are used in a range of domestic products. Alkalis help to remove greasy dirt.

Key words

antacids
alkalis
corrosive
irritant

Question

1 Milk of Magnesia contains magnesium hydroxide. Write a word equation to show what salt will form in your stomach if you take this antacid medicine.

Neutralisation

Alkalis neutralise acids to form salts. The word equation is:

metal hydroxide + acid → salt + water

Hazards of acids and alkalis

Concentrated acids are extremely hazardous. They are **corrosive**. Dilute solutions of acids in water are much less hazardous. Dilute hydrochloric acid is often used in solutions that are so dilute that they are not even classified as **irritant**.

Dilute sulfuric acid is an irritant. Nitric acid is more hazardous. Even quite dilute solutions can be corrosive.

Some alkalis attack skin and flesh. Examples are sodium hydroxide and potassium hydroxide. They used to be called caustic alkalis because of the way they attack living tissue. Even dilute solutions of these compounds can be hazardous, especially in the eyes. They are corrosive.

If you rub a small drop of dilute sodium hydroxide solution between the tips of two fingers they will soon feel soapy as the alkali attacks the grease on the skin.

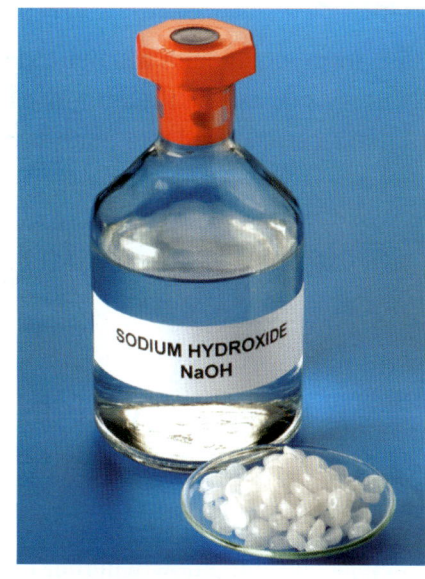

Pure sodium hydroxide is a white solid. It is soluble in water and used in solution as a laboratory alkali. It is corrosive.

HARMFUL
a substance which if inhaled, ingested, or taken in through the skin, may involve limited health risks

TOXIC
a substance which if inhaled, ingested, or taken in through the skin, may involve serious acute or chronic health risks and even death

OXIDISING
a substance which produces a reaction giving off great heat when in contact with other substances, particularly flammable substances

IRRITANT
a non-corrosive substance which through immediate, prolonged, or repeated contact with the skin or eyes, may cause inflammation or lesions

CORROSIVE
a substance which on contact with living tissues may destroy the tissues

HIGHLY FLAMMABLE
a liquid which may easily catch fire or a solid which burns after brief contact with a flame or which gives off flammable gases in contact with water

Substances harmful to health should display recognised warning symbols. These are some of the hazard symbols used on common chemicals.

Salts

Salts form when acids are neutralised by a metal oxide or hydroxide. So every salt can be thought of as having two parents. Salts are related to a parent acid and to a parent metal oxide or hydroxide.

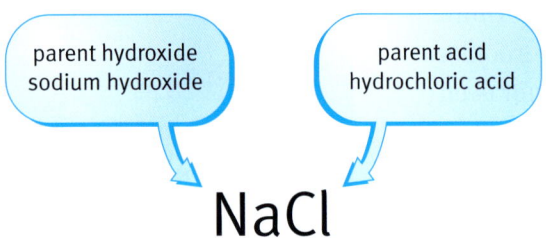

parent hydroxide
sodium hydroxide

parent acid
hydrochloric acid

NaCl

Soluble or insoluble?

Chemists need to know whether or not a salt is **soluble** in water before making it or using it. You can look up the solubility of a chemical in data tables, but it is useful to be familiar with the patterns of solubility of common salts. A solution of a chemical in water is an **aqueous** solution.

Salt	Soluble	Insoluble
nitrates	all soluble	none
chlorides	mostly soluble	silver and lead chlorides
sulfates	mostly soluble	barium and lead sulfates; calcium sulfate is very slightly soluble
carbonates	sodium and potassium carbonates	mostly insoluble

Salt crystals often form by **crystallisation** from a solution in water. When this happens some water molecules may become part of the crystal structure. Chemists call this water of crystallisation. They include the water in the formula of the salt. Examples are blue copper sulfate, $CuSO_4.5H_2O$, and hydrated sodium carbonate, $Na_2CO_3.10H_2O$.

Why make salts?

Salts have many uses. They help to keep us clean and healthy. They decorate our lives. They protect and preserve crops and food.

Key words

soluble
aqueous
crystallisation

Questions

Look at page 59 for help with the chemical formulae.

1 Give the chemical formula for potassium nitrate and suggest some ways that it is useful.

2 Give the name and chemical formula of two salts that are insoluble in water.

3 Give the name and chemical formula of two salts that can form aqueous solutions.

Lead chromate is the bright yellow salt used by Van Gogh to paint this picture. He was fond of this colour and also used it to paint the frame.

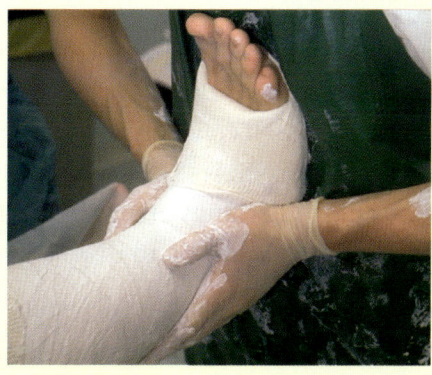

Calcium sulfate, $CaSO_4$, exists as the mineral anhydrite or in a hydrated form, $CaSO_4.2H_2O$, as gypsum. Heating gypsum drives off part of the water in the crystals, converting the mineral to plaster of Paris. Plaster sets hard when mixed with water. It is used to make casts for broken bones and plasterboard for buildings.

Potassium nitrate is used in curing meat to make things like bacon and pastrami. It is also used as fertiliser and as an important constituent of gunpowder and fireworks.

Sodium benzoate is widely used as a food preservative. It is used to prevent bacteria spoiling food. It works best if the pH is low, so it is added to foods such as jams, salad dressing, fruit juices, pickles, and carbonated drinks.

Sodium citrate is a food additive that helps to control the pH of foods and make antioxidants more effective. So it prevents food going off in the presence of air.

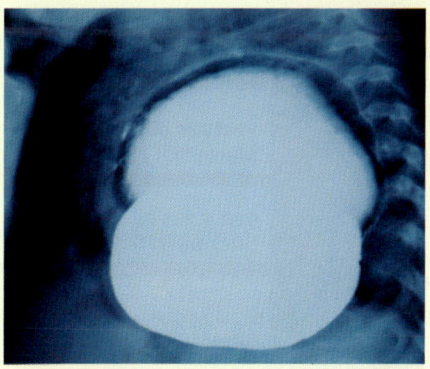

Barium sulfate is used in hospitals to investigate the gut. The patient is given a 'meal' of barium sulfate and then photographed with X-rays. Problems in the gut show up on the film. This 9-month-old child is suffering from a hernia. Part of the stomach has been pushed into the chest cavity.

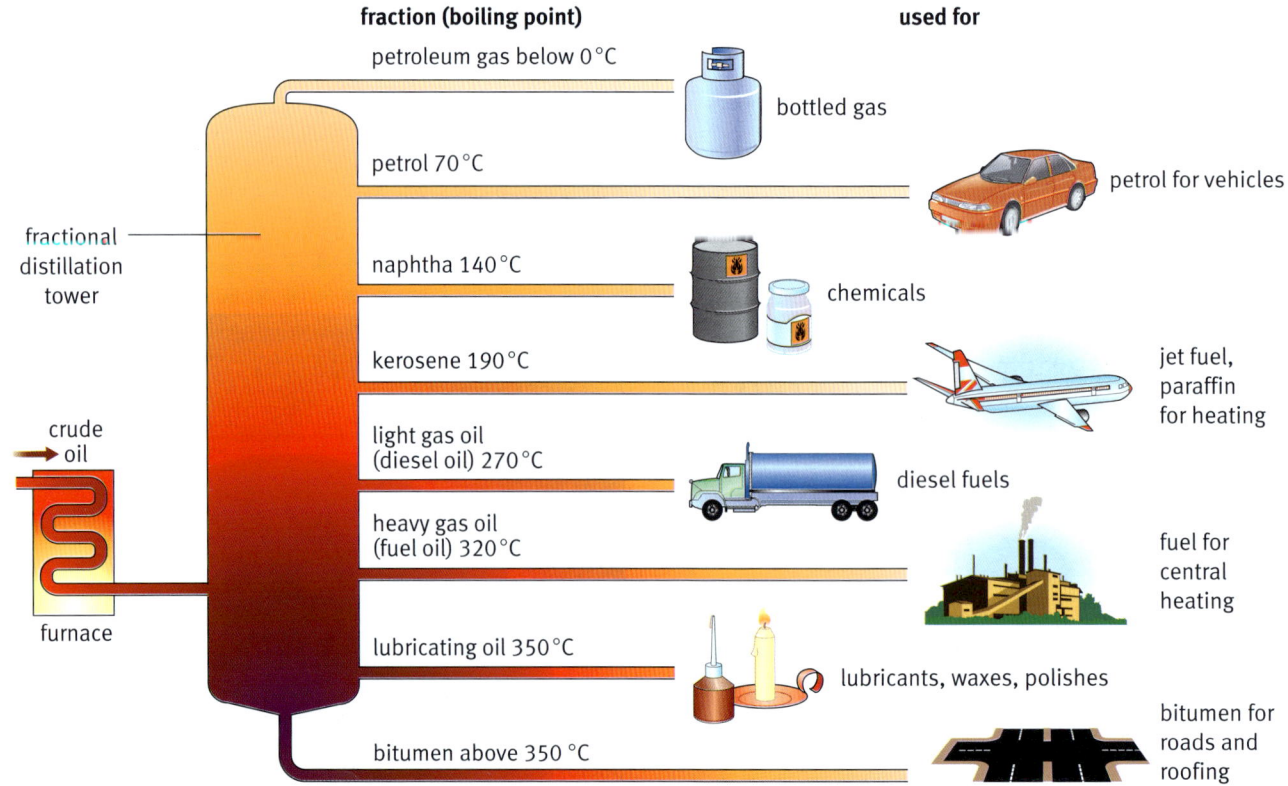

Organic chemicals

Carbon is a very unusual element. Its atoms can join up in many different ways to make chains and rings. Carbon forms so many compounds that it gives rise to its own chemistry. Chemists call this **organic chemistry** because carbon is the element of life. All the important compounds in living things are carbon compounds. This includes carbohydrates, proteins, fats, and DNA.

Crude oil, coal, and natural gas are fossil fuels. They are also sources of **petrochemicals**. Petrochemicals are carbon compounds. We need them to make plastics, fibres, medicines, dyes, and detergents.

The chemicals in oil formed under the sea from the decaying remains of tiny sea creatures. This happened millions of years ago. This means crude oil is a chemical fossil, hence the term 'fossil fuel'. It also helps to explain why chemists call these compounds **organic chemicals** – 'organic' means 'from living things'.

An oil refinery separates oil into **fractions** by **distillation**. Each fraction contains organic compounds with about the same boiling point. Fractions flow continuously from the column. After separating the fractions, the refinery carries out further processing to make products that other industries want to buy.

> **Key words**
> organic chemistry
> petrochemicals
> organic chemicals
> fractions
> distillation

fraction (boiling point)   **used for**

petroleum gas below 0 °C — bottled gas

petrol 70 °C — petrol for vehicles

fractional distillation tower

naphtha 140 °C — chemicals

kerosene 190 °C — jet fuel, paraffin for heating

crude oil

light gas oil (diesel oil) 270 °C — diesel fuels

heavy gas oil (fuel oil) 320 °C — fuel for central heating

furnace

lubricating oil 350 °C — lubricants, waxes, polishes

bitumen above 350 °C — bitumen for roads and roofing

The tower separates the mixture that is crude oil into fractions, depending on the boiling points of the compounds in the mixture.

Hydrocarbons

Hydrocarbons are organic chemicals made up of carbon and hydrogen only. Many of them come from oil and natural gas. The main chemical reaction of hydrocarbons is burning. They burn well when they react with oxygen in the air. This makes them good fuels.

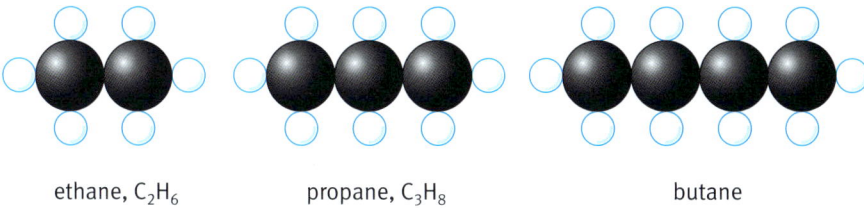

ethane, C_2H_6 propane, C_3H_8 butane

Three hydrocarbon molecules

The smallest hydrocarbon molecules are found in the fractions of crude oil with low boiling points. The fractions with very high boiling points, which are thick and slimy, contain much larger hydrocarbon molecules. Petrol for cars is a mixture of smaller hydrocarbons. It is specially formulated to suit the local climate, as it is important that it vaporises easily in the car engine.

The mixture of hydrocarbons in petrol is formulated to suit the climate.

Key words
hydrocarbons

Questions

1 What is the chemical formula of butane?

2 Draw a diagram of the pentane molecule C_5H_{12}.

3 Draw a diagram to show what happens to methane, CH_4, when it is burnt in oxygen, O_2, to form carbon dioxide, CO_2, and water, H_2O.

4 In which of the crude oil fractions would you expect these hydrocarbons to be found?

   a nonane, boiling point 150 °C

   b propane, boiling point −42 °C

   c hexane, boiling point 69 °C

   d eicosane, boiling point 344 °C

Alcohols

Ethanol is the best known member of the family of **alcohols**. It is the alcohol in beer, wines, and spirits. Ethanol is also a very useful solvent, especially when water is not suitable. Ethanol is an example of a **non-aqueous solvent**. Ethanol **evaporates** quickly, which makes it a good solvent for perfumes and aftershave lotions.

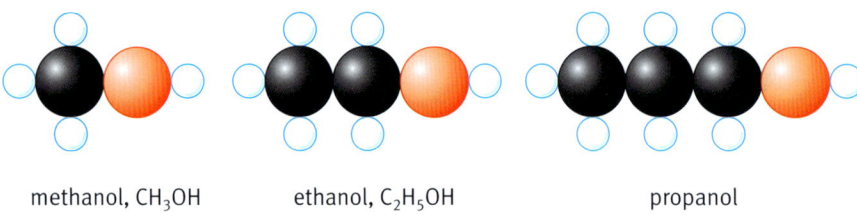

methanol, CH_3OH ethanol, C_2H_5OH propanol

Three alcohols. Note that all three have OH as part of the molecule.

The names of organic compounds are based on the number of carbon atoms in the molecules. All 'eth' compounds have two carbon atoms in their molecules. A molecule of ethanol is almost the same as a molecule of ethane, except that one of the hydrogen atoms has been replaced by OH. It is the OH which gives alcohols their special properties. The OH is the **functional group** of an alcohol.

Carboxylic acids

Carboxylic acids make up another family of carbon compounds. These acids contain the COOH functional group. The most familiar of these acids is ethanoic acid, the acid in vinegar.

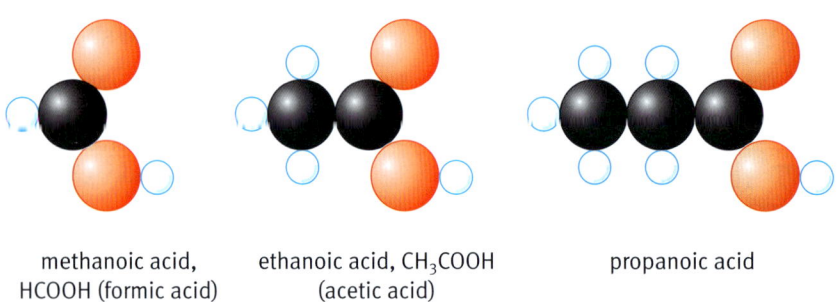

methanoic acid,   ethanoic acid, CH_3COOH propanoic acid
HCOOH (formic acid)   (acetic acid)

Three carboxylic acids. Note that all three have COOH as part of the molecule.

> ### Questions
> 1 What is the chemical formula for propanoic acid?
> 2 Classify these organic compounds under three headings, Hydrocarbon, Alcohol, Carboxylic acid:
>
> C_4H_9OH C_2H_4 $C_5H_{11}COOH$ C_6H_6 HCOOH
> methanol   ethane   ethanoic acid   butanol

Key words

alcohols
non-aqueous solvent
evaporates
functional group
carboxylic acids

This red wood ant can spray attackers with concentrated methanoic acid. The traditional name for the acid is 'formic acid' based on the Latin name for an ant.

Esters

Carboxylic acids react with alcohols to form compounds called **esters**:

a carboxylic acid + an alcohol → an ester + water

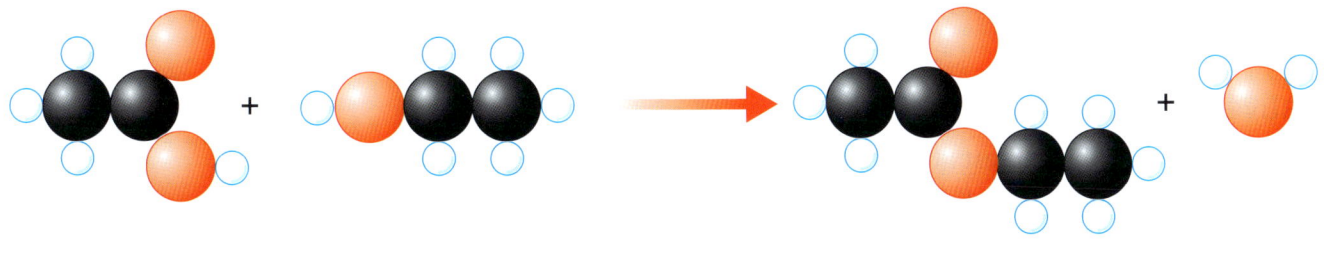

ethanoic acid, CH$_3$COOH   ethanol, C$_2$H$_5$OH   ethyl ethanoate, CH$_3$COOC$_2$H$_5$ water, H$_2$O
(acetic acid)

Ethanol reacts with ethanoic acid to form the ester ethyl ethanoate and water.

Esters are very common. When you eat a banana, suck a fruit sweet, or remove nail varnish with a solvent, you smell the strong fruity smell of an ester. In a kilogram of ripe pineapple there is about 120 mg of the ester ethyl ethanoate, along with smaller amounts of other esters which together make up the flavour of the fruit.

Esters are used in artificial perfumes for their sweet smell. They are also used as artificial flavourings in foods such as ice cream, yogurt, soft drinks, and sweets. They are non-aqueous solvents, used in making fats, paint, and varnishes. Butter, margarine, and sunflower oil are all mixtures of esters. Polyester is a synthetic fibre used for clothing and other fabrics. The same polyester is used to make the bottles for carbonated drinks.

> **Key word**
> esters

Esters are used to make all these products.

> **Questions**
>
> 1 List five different consumer products that have a 'fruity' taste or smell.
>
> 2 Explain why nail varnish does not come off in water, then explain why it can be removed with nail varnish remover. Use the words soluble, insoluble, and non-aqueous solvent in your answer.
>
> 3 Copy and complete the word equations for the formation of these esters:
>
>   a methanol + ethanoic acid →
>   methyl ethanoate + _____
>
>   b ethanol + butanoic acid →
>   ethyl _____ + water
>
>   c pentanol + _____ acid →
>   pentyl ethanoate + water

Chemical quantities

Chemists need to be able to work out how much of each reactant to mix together to make the quantity of the product they need. To do this they need to be able to turn the symbols in the balanced chemical equation into masses in grams. This is possible given the relative masses of the atoms.

Relative atomic masses

A hydrogen atom is the lightest atom. On the scale used by chemists, hydrogen has a relative mass of 1. All other atoms have more mass. The **relative atomic mass** of carbon is 12. This means that a carbon atom is twelve times heavier than a hydrogen atom.

The relative atomic mass of magnesium is 24, so magnesium atoms are twice as heavy as carbon atoms and 24 times as heavy as hydrogen atoms.

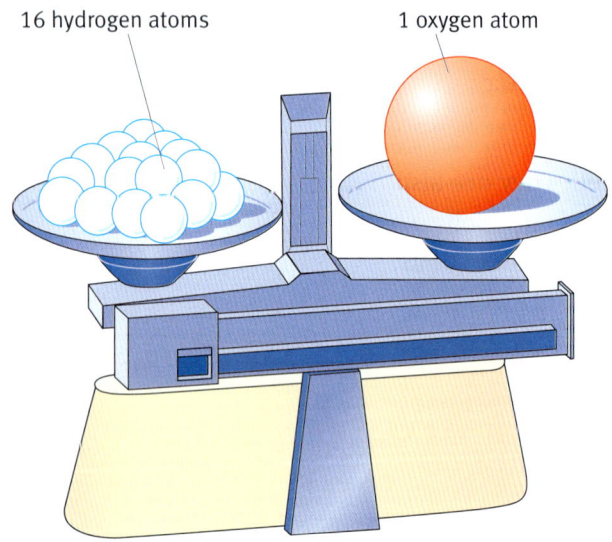

16 hydrogen atoms   1 oxygen atom

One oxygen atom has the same mass as 16 hydrogen atoms. The relative atomic mass of oxygen is 16.

Relative formula masses

Adding up the relative atomic masses for all the atoms in the formula of a compound gives the **relative formula mass**.

Example 1: The relative formula mass of calcium carbonate

The formula of calcium carbonate is: $CaCO_3$

The relative atomic masses are: Ca = 40, C = 12, and O = 16

The relative formula mass of the compound = $40 + 12 + (3 \times 16) = 100$

(Note that these are relative masses so you do not include units here.)

Reacting masses

Given the relative atomic and formula masses, it is possible to work out the **reacting masses** from a balanced equation.

> ## Rules for working out reacting masses
>
> **STEP 1** Write down the balanced symbol equation.
>
> **STEP 2** Work out the relative formula mass of each reactant and product.
>
> **STEP 3** Write the relative reacting masses under the balanced equation, taking into account the numbers used to balance the equation.
>
> **STEP 4** Convert to reacting masses by adding the units (g, kg, or tonnes).
>
> **STEP 5** Scale the quantities to amounts actually used in the synthesis or experiment.

Follow these rules for calculating reacting masses.

Example 2: What are the reacting masses when sulfuric acid reacts with sodium hydroxide?

Step 1

$$2NaOH + H_2SO_4 \rightarrow Na_2SO_4 + 2H_2O$$

Step 2

The relative atomic masses are: Na = 23, H = 1, S = 32, and O = 16

Relative formula mass of NaOH = 23 + 16 + 1 = 40

Relative formula mass of H_2SO_4 = (2 × 1) + 32 + (4 × 16) = 98

Relative formula mass of Na_2SO_4 = (2 × 23) + 32 + (4 × 16) = 142

Relative formula mass of H_2O = (2 × 1) + 16 = 18

Steps 3 and 4

$2NaOH$	+	H_2SO_4	$\rightarrow$	Na_2SO_4	+	$2H_2O$
2 × 40 = 80		98		142		2 × 18 = 36
80 g		98 g		142 g		36 g

The reacting masses have the same number value in grams as the relative masses. Note that the proportions stay the same even working on a different scale. So 80 tonnes of sodium hydroxide react with 98 tonnes of sulfuric acid to make 142 tonnes of sodium sulfate and 36 tonnes of water.

This is another way of showing that the equation is balanced. No atoms are created or destroyed and so (80g + 98g) = 178g of reactants gives (142g + 36g) = 178g of products. The masses of reactants and products are equal.

Question

1 Use these relative atomic masses to work out the reacting masses for the reactions:
H = 1, C = 12,
N = 14, O = 16,
Na = 23, S = 32,
Cl = 35.5, Ca = 40,
Cu = 64.

a NaOH + HCl →
   NaCl + H_2O

b CuO + H_2SO_4 →
   $CuSO_4$ + H_2O

c $CaCO_3$ + 2HNO$_3$ →
   $Ca(NO_3)_2$ + CO_2 + H_2O

Key words

relative atomic mass
relative formula mass
reacting masses

Yields

The **yield** is the quantity of product obtained from known amounts of starting materials. The quantities used in a chemical synthesis are often different from those in the balanced equation, so we scale the figures up or down.

Actual yield and theoretical yield

The **actual yield** is the mass of product after separating it from the mixture, purifying, and drying it.

The **theoretical yield** is the mass of product expected if the reaction goes exactly as shown in the balanced equation, with no by-products and no losses while transferring chemicals from one container to another. The actual yield is always less than the theoretical yield.

Example 1: The theoretical yield of silver chloride made from 6.80 g of silver nitrate

Silver chloride precipitates as a solid on mixing solutions of silver nitrate and sodium chloride. In this example it is only necessary to work out the reacting masses of the silver nitrate and the silver chloride. Sodium chloride is cheap and added in excess.

silver nitrate + sodium chloride → silver chloride + sodium nitrate

$$AgNO_3 \quad + \quad NaCl \quad \rightarrow \quad AgCl \quad + \quad NaNO_3$$
$$170 \qquad\qquad\qquad\qquad\qquad 143.5$$

This shows that, theoretically, 170g of silver nitrate should give 143.5g of silver chloride. In the lab we would use a much smaller quantity of silver nitrate, only a fraction of the reacting mass in the equation. The same fraction of the reacting mass of silver chloride forms.

So, if we use 6.80 g of silver nitrate this should give $\dfrac{6.80\,g}{170\,g} \times 143.5\,g = 5.74\,g.$

Percentage yield

The **percentage yield** is the percentage of the theoretical yield that is actually obtained. It is always less than 100%. It shows how well the reaction worked and how much product is lost during its collection.

Example 2: What is the percentage yield if 6.80 g of silver nitrate produces 3.87 g of silver chloride?

theoretical yield = 5.74 g (see example 1)

actual yield = 3.87 g

percentage yield = $\dfrac{\text{actual yield}}{\text{theoretical yield}} \times 100 = \dfrac{3.87\,g}{5.74\,g} \times 100 = 67\%$

Questions

1 A preparation of sodium sulfate began with 8.0 g of sodium hydroxide. The theoretical yield of sodium sulfate from this amount of sodium hydroxide is 14.2 g. The actual yield was 12.0 g. Calculate the percentage yield.

2 What is the theoretical yield of zinc sulfate when 12.5 g of zinc carbonate, $ZnCO_3$, reacts with excess sulfuric acid? (Relative atomic masses: C = 12, O = 16, S = 32, Zn = 65)

Key words

yield
actual yield
theoretical yield
percentage yield

Purity of chemicals

Suppliers of chemicals offer a range of grades of chemicals. In a school laboratory you might use technical, general laboratory, and analytical grades.

Calcium carbonate, for example, is used in a blast furnace to extract iron from its ores. It is also an ingredient of indigestion tablets. The iron industry can use limestone straight from a quarry. Limestone has some impurities, but they do not stop it from doing its job in a blast furnace.

Indigestion tablets have to be very **pure**. The standards of purity for chemicals in medicines are set out in the *British Pharmacopoeia* (BP). Calcium carbonate purified to BP quality is safe to swallow.

Purifying a chemical is done in stages. Each stage takes time and money, and becomes more difficult. So the higher the purity, the more expensive the chemical. Manufacturers therefore buy the quality most suitable for their purpose.

The science

Questions

1 Make a list of all the different uses of sodium chloride (salt) that you know. Beside each one write which grade of salt you think it would be best to use.

2 Explain why pure salt for food use is cheaper than sodium chloride general purpose reagent.

Key word

pure

Industrial chemicals			Laboratory chemicals		
Calcium carbonate			*Calcium carbonate*		
limestone	£0.01 per kg (= £10 per tonne)		general purpose reagent	£3 per kg	
BP quality (for medicines)	£0.50 per kg (= £500 per tonne)		analytical grade	£30 per kg	
Sodium chloride			*Sodium chloride*		
rock salt	£0.05 per kg (= £50 per tonne)		general purpose reagent	£1.50 per kg	
pure salt for food use	£0.07 per kg (= £70 per tonne)		analytical grade	£9 per kg	

Approximate prices in 2003. Laboratory chemicals, used in small amounts, are much more expensive than industrial chemicals, bought and sold by the tonne. For both types, price increases with purity.

When deciding what grade of chemical to use it is important to know

- the amount of impurities

- what the impurities are

- how they can affect the process

- whether they will end up in the product, and whether it matters if they do

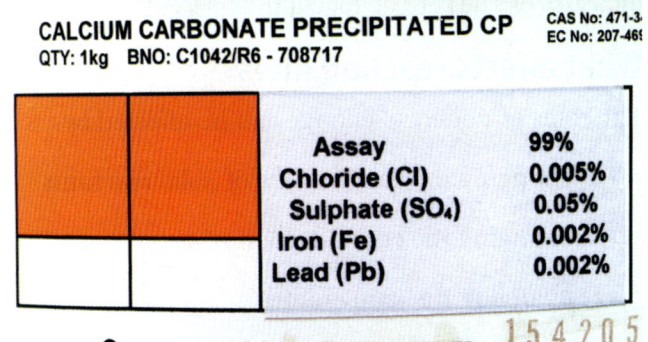

CALCIUM CARBONATE PRECIPITATED CP
CAS No: 471-3
EC No: 207-46
QTY: 1kg   BNO: C1042/R6 - 708717

Assay	99%
Chloride (Cl)	0.005%
Sulphate (SO₄)	0.05%
Iron (Fe)	0.002%
Lead (Pb)	0.002%

154205

The label on a laboratory bottle of calcium carbonate

Controlling reaction rates

Some chemical reactions seem to happen in an instant. An explosion is an example of a very fast reaction.

Other reactions take time – seconds, minutes, hours, or even years. Rusting is a slow reaction and so is the rotting of food.

Rate of reaction

Your pulse rate is the number of times your heart beats every minute. The production rate in a factory is a measure of how many articles are made in a particular time. Similar ideas apply to chemical reactions. Chemists measure the **rate of a reaction** by finding the quantity of product produced or the quantity of reactant used up in a fixed time.

An explosion is an example of a very fast chemical reaction, as this fireball from a detonation of gunpowder shows.

For the reaction

$$Mg(s) \; + \; 2HCl(aq) \; \rightarrow \; MgCl_2(aq) \; + \; H_2(g)$$

the rate can be found quite easily by measuring either the disappearance of the magnesium or the formation of hydrogen gas

In most chemical reactions the rate changes with time. The graph shows the volume of hydrogen formed against time for the reaction above. The graph is steepest at the start, showing that the rate of reaction was greatest at that point. As the reaction continues the rate decreases until the reaction finally stops. The steepness of the line is a measure of the rate of reaction.

What affects reaction rates?

The rates of chemical change can be affected by

- changing the surface area of solid reactants
- changing the **concentration** of reactants in solution
- changing the temperature
- adding a catalyst

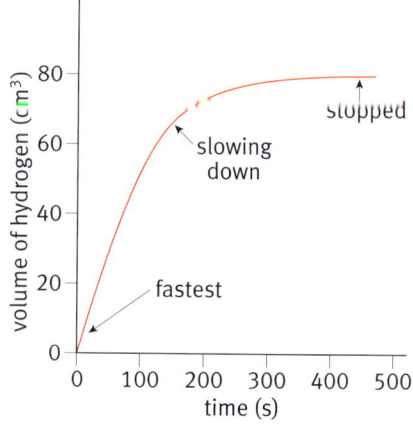

A plot of the volume of hydrogen formed against time for a reaction of magnesium with hydrochloric acid

Explaining reaction rates

Atoms and molecules can only react if they bump into each other. The more they collide and the harder they bump into each other, the faster the reaction is likely to go.

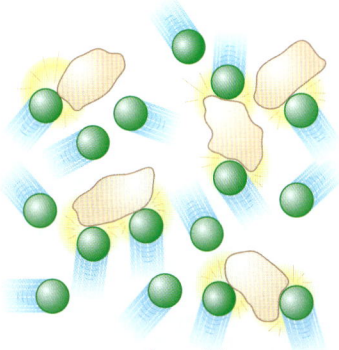
several small lumps

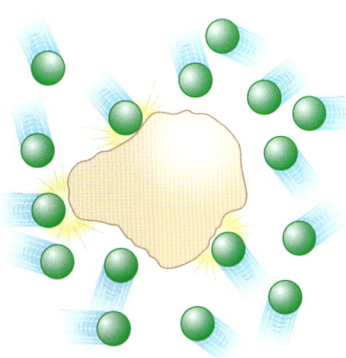
one big lump

Grinding up a solid into smaller pieces increases its surface area. If the surface area is bigger, more molecules can collide with the solid and react. So the smaller the solid particles, the bigger the surface area and the faster the reaction.

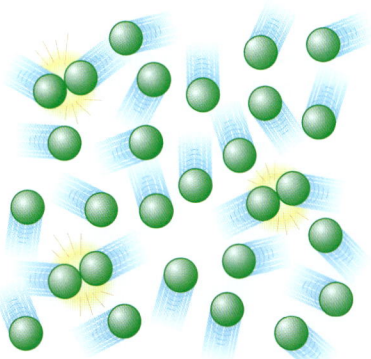
higher concentration

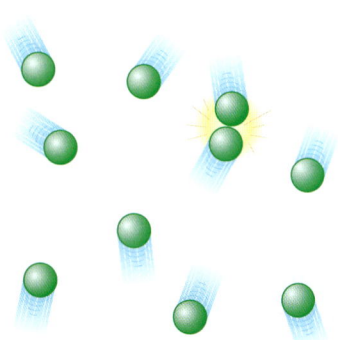
lower concentration

If the reactants are more concentrated, the atoms or molecules are closer together. They collide more often and the reaction goes faster. The higher the concentration, the faster the reaction.

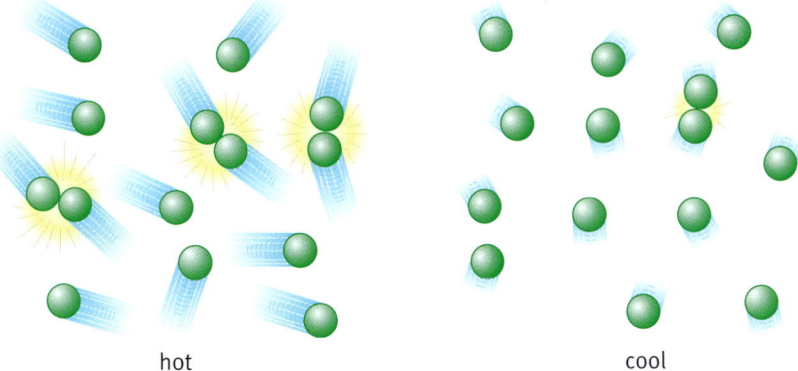
hot   cool

The hotter the reactants, the faster their atoms or molecules collide. When hotter they collide more often and with more energy. The higher the temperature, the faster the reaction.

Questions

1 How would you account for the fact that

  a sliced bread goes stale more quickly than unsliced bread

  b there is a danger of explosions in flour mills?

2 How is it possible to control conditions to slow down or stop these changes?

  a the rusting of iron

  b a chip-pan fire

  c milk going sour

3 How is it possible to control conditions to speed up these changes?

  a the setting of an epoxy glue

  b the cooking of an egg

  c the conversion of oxides of nitrogen in car exhausts to nitrogen

  d the conversion of sugar to alcohol and carbon dioxide

Key words

rate of reaction
concentration

Catalysts

What is a catalyst?

A **catalyst** is a chemical that speeds up a chemical reaction. It takes part in the reaction, but is not used up. At the end of the reaction there is the same quantity of catalyst as there was at the beginning.

In the exhaust pipes of petrol vehicles is a platinum/rhodium catalyst which helps to control air pollution. The solid catalyst is very finely divided, which means it is in very tiny pieces which are spread thinly over a large surface area. The job of the catalyst is to speed up reactions that convert oxides of nitrogen and hydrocarbons into chemicals that are less harmful.

Catalysts in industry

Catalysts are essential in many industrial processes. The whole of the petrochemical industry depends on the use of catalysts. They make many processes more cost effective.

Modern catalysts can be highly selective. A suitable catalyst can make it possible to speed up the reaction that gives the right product, but not the ones that create unwanted by-products. Research into better catalysts is an important area of scientific work.

Enzymes – natural catalysts

Enzymes are naturally occurring biological compounds. They control the speed of reactions taking place in living organisms. Chemists are now harnessing these naturally occurring enzymes and using them to make chemicals in a more environmentally friendly way.

Enzymes are only needed in very small amounts, they are very selective, and they work at lower temperatures than inorganic catalysts. The right enzyme makes it possible to use much milder reaction conditions. This makes savings in energy, and in the cost of building and maintaining expensive plant.

The manufacture of ethanoic acid from methanol and carbon monoxide is only possible in the presence of a catalyst.

methanol + carbon monoxide → ethanoic acid

$$CH_3OH(g) + CO(g) \rightarrow CH_3COOH(g)$$

Reaction conditions: pressure 30 atmospheres; temperature 200 °C; catalyst iridium

Key words

catalyst
enzymes
synthetic routes

Choosing a synthetic route

Often there is more than one way to make a chemical. Scientists have a choice of **synthetic routes** and must decide which one to choose.

Ethanol, for example, can be made by fermenting sugars or starch. Alternatively it can be manufactured from ethene, which is a petrochemical.

Distilling alcohol from the products of fermentation at a Tequila distillery in Mexico

What is the optimum route? How does a synthetic chemist decide which route to choose? There are a number of factors to consider, including:

- the availability and cost of starting materials – the materials should be easy to obtain, cheap, and renewable if possible

- the energy requirements and their costs – the route should need as little energy input as possible

- the efficiency of the route – the route should give a high yield of the required product and the minimum of by-products

- the hazards associated with the chemicals and the process – where possible the reacting chemicals should not be toxic, highly flammable, or corrosive, and the process should not be hazardous to the operators

- the disposal or recycling of by-products – ideally by-products should be recycled or, if this is not possible, be disposed of easily, safely, and cheaply

Often there is a conflict. For example, one route might use the cheapest starting materials but give poor yields. Another route might give high yields but require a high energy input.

Scaling up

Laboratory scale preparation of zinc chelate

Chemists make zinc chelate from two solids (zinc sulfate and EDTA) and a solution of an alkali in water (sodium hydroxide). The starting chemicals come from bottles on the laboratory shelves; the quantities are small and easily transferred. Technical grade zinc sulfate can be used as long as the impurities are removed later by filtering.

The reaction

▶ Heat the EDTA and water on a hotplate while stirring with a glass rod.

▶ Add the zinc sulfate and stir.

▶ Add the sodium hydroxide solution and continue stirring.

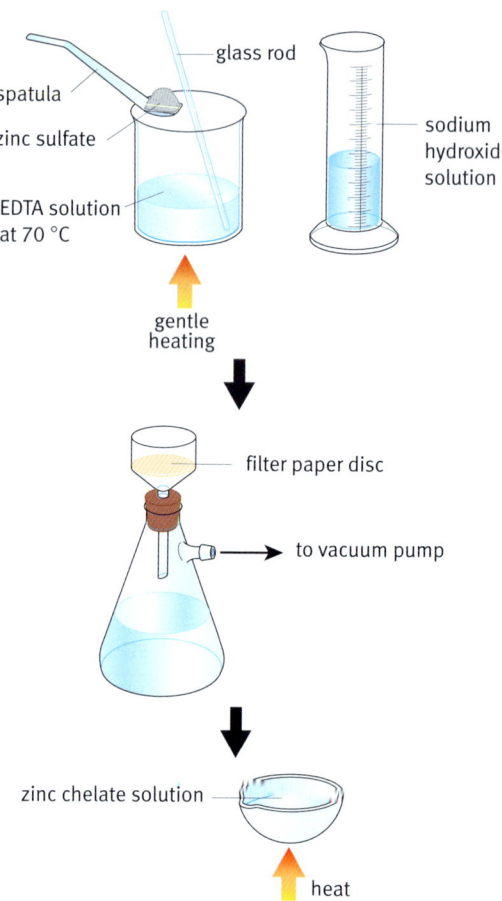

Filtration

▶ Pour the hot solution into the filter funnel lined with filter paper.

▶ Reduce the pressure in the flask to speed up the filtration. Collect the filtrate.

Separation of the solid product

▶ Evaporate the filtrate to concentrate the solution.

▶ Set aside to cool and crystallise. Weigh the dry product.

What is zinc chelate?

Traces of zinc are essential for healthy plant growth. Farmers can improve yields of some crops by spraying the leaves with the soluble zinc compound called zinc chelate.

Industrial scale preparation of zinc chelate

Scaling up from a laboratory to an industrial process does not change the chemistry but it does change the way the operations are carried out. The EDTA and zinc sulfate are bought in 25 kg bags; the water comes from the mains, while sodium hydroxide solution flows in a pipe from a large storage tank.

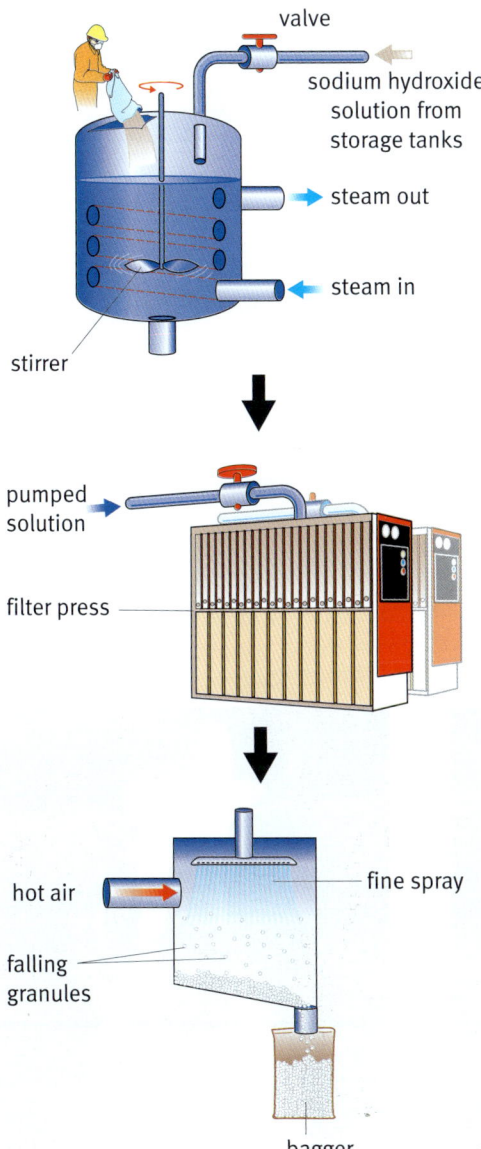

The reaction

▶ Open the valve to let water be pumped into the reaction vessel. Switch on the large motor-driven stirrer, open the valve to allow steam to flow through the pipes, and heat the reaction vessel. Tip in 40 bags of EDTA.

▶ Add 29 bags of zinc sulfate.

▶ Open the valve to run in sodium hydroxide solution. Stop the flow of steam because the reaction is exothermic and the mixture must not get too hot.

Filtration

▶ Pump the hot solution from the reaction vessel into a filter press with cloth filters.

▶ Use pressure to force the solution through the filters. Collect the filtrate in a heated storage tank.

Separation of the solid product

▶ Spray the hot solution into the drying chamber. As the droplets fall through the blast of hot air from the gas burner, the water evaporates.

▶ Collect the dry granules from the bottom of the drying chamber and pack in 25 kg bags.

Batch or continuous?

Demand for zinc chelate is seasonal. It is only needed when the crops are growing, so the manufacturer makes it by a **batch process**. The reaction vessel, filter, and drying chamber are used to make other salts when demand for zinc chelate is low.

A **continuous process** could be used if the product was sold all over the world, so that the demand was more constant. A lot of money is needed to set up an automated continuous process, but once installed, it can be kept running continuously with lower labour costs and higher productivity.

Emulsions and suspensions

Try to imagine what the world would be like with no clouds, no brilliant sunsets, and no fog on city streets. Think how different your diet would be with no butter or milk, no ice cream or jellies, no ketchup, salad cream, bread, or cake. Consider living without cosmetics or medical creams and no paints.

What all these things have in common is that they consist of one chemical very finely dispersed (spread) in another. Milk is a good example. It consists of droplets of fat finely dispersed in a watery solution.

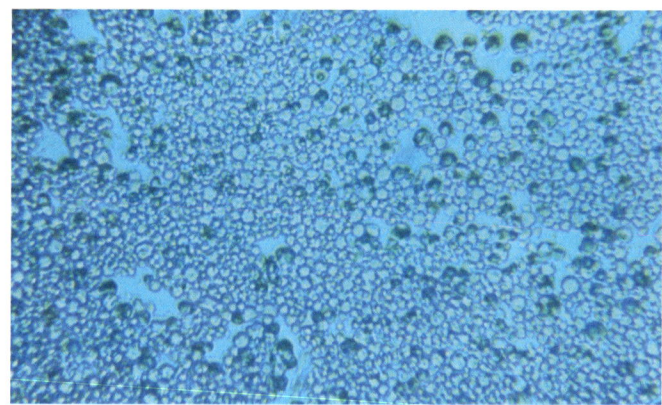

False colour image of fat droplets in milk under the microscope (magnification × 200)

Emulsions

Milk is an example of an **emulsion**. It consists of tiny droplets of an oily liquid dispersed in a watery liquid.

Many emulsions tend to separate into two layers unless an **emulsifying agent** is present in the mixture. You can see this in a salad dressing made with oil and vinegar. Shaking the two liquids together breaks up the oil into droplets. But if you leave the dressing to stand, the oil droplets soon join up to form bigger drops and this continues until the two liquids separate into two layers again.

> **Key words**
> emulsion
> emulsifying agent
> stabilisers
> suspension

The oil and vinegar in this bottle have not been shaken.

The bottle has been shaken, creating a temporary emulsion of oil and vinegar.

The shaken bottle has been allowed to stand; the two layers of oil and vinegar have begun to re-form.

Cooks use various emulsifying agents to make an emulsion of oil and vinegar more permanent. To make mayonnaise they use egg yolk. Egg yolk contains lecithin which is a natural emulsifier. Food manufacturers usually use lecithin from soya bean oil because it is cheaper. Lecithin is one example of a food additive.

Stabilisers can also help to make permanent food emulsions. They increase the thickness of the mixture which stops the oil drops coming together. Natural polymers used to stabilise emulsions in this way include starch and extracts from seaweeds called agar and carrageenan.

Many cosmetics are emulsions. These are of two types: oil-in-water and water-in-oil emulsions.

Questions

1 Why do cooks need emulsifying agents?

2 Give examples of two emulsifying agents used to make mayonnaise.

3 Why do some food products include stabilisers as well as an emulsifying agent?

4 What is the difference between an emulsion and a suspension?

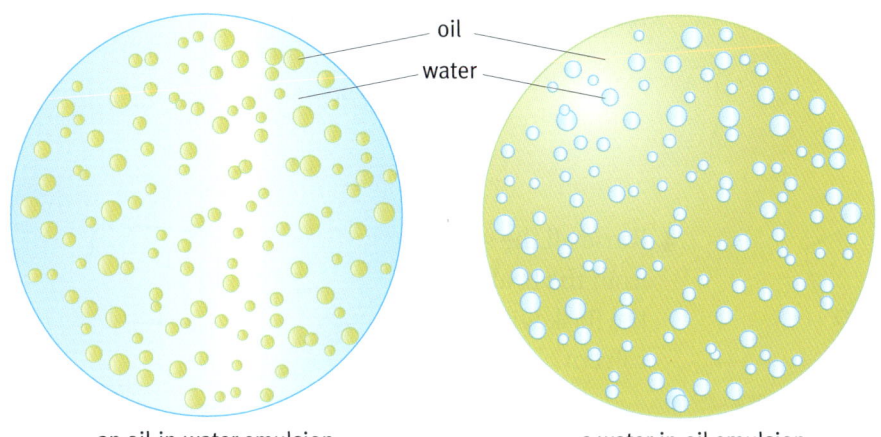

oil

water

an oil-in-water emulsion   a water-in-oil emulsion

Suspensions

A soluble solid dissolves in a liquid to give a solution. Colourless or coloured, all solutions are transparent.

An insoluble salt does not dissolve. Shake or stir an insoluble solid with a liquid and it forms a **suspension**. A suspension consists of small specks of a solid dispersed in a liquid. Suspensions are cloudy. They may be quite runny, like Milk of Magnesia, or thick like toothpaste.

The insoluble solid in a suspension may settle out quickly. Other suspensions last longer and may take a very long time to separate.

Solutions are transparent. Suspensions are not. This person can see through the solution, but not through the suspension.

Preparing a solution with a known concentration

→ Principle

A solution consists of a **solute** (usually a solid) dissolved in a **solvent** (often water).

It is important to be able to make up a solution with a **concentration** that is accurately known.

For units of concentration, see page 59.

→ Procedure

1 Accurately weigh the solid.

2 Dissolve the solute in a small amount of solvent, warming if necessary.

stirring rod

3 Transfer the solution to a graduated flask, letting the liquid run down a glass rod to make sure it does not spill.

stirring rod

paper wedge

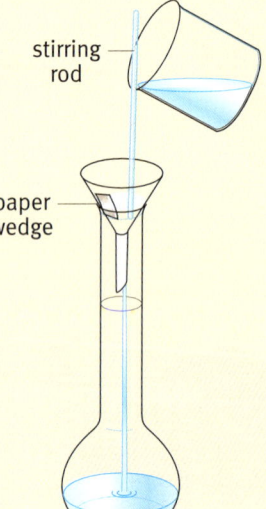

4 Rinse all the solution from the beaker into the flask with more solvent.

wash bottle

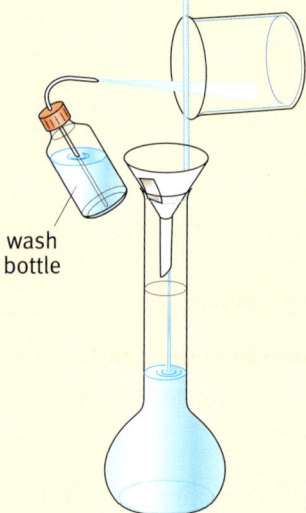

5 Carefully add solvent up to the mark on the flask, drop by drop.

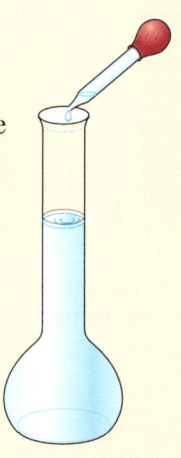

6 Stopper and mix well.

7 Pour the solution into a clean, dry container labelled to show the type of solution, the date the solution was made, and its concentration (in grams per litre).

NaCl solution 20 g/l

Making an insoluble salt

CHECK SAFETY
Never work
unsupervised

→ Principles

Two solutions of soluble salts may react to produce one **insoluble** salt and a new soluble salt. The insoluble salt separates as a solid **precipitate**. The new soluble salt stays in solution.

soluble	+	soluble	→	insoluble	+	soluble
salt A		salt B		salt		salt C
(aq)		(aq)		(s)		(aq)

→ Procedure

1 Mix the solutions of two soluble salts. A precipitate of the insoluble salt forms immediately.

salt solution **A**

salt solution **B**

2 Filter off the precipitate.

insoluble salt residue

filtrate is salt solution **C**

3 Wash the precipitate on the filter paper with pure water. This removes the soluble salts.

water

4 Open the filter paper and leave the precipitate to dry at room temperature (or in an oven).

filter paper

5 Scrape the dry solid into a weighed sample tube. Reweigh. Label the container to show the name of the solid and the date.

name of salt

Making soluble salts

→ Principle

These three general reactions of acids can be used to make **soluble** salts.

acid + metal → salt + hydrogen

acid + metal oxide or hydroxide → salt + water

acid + carbonate → salt + carbon dioxide + water

The procedure below is used to make salts where the metal does not react with water, or the metal oxide, metal hydroxide, or carbonate does not dissolve in water.

→ Procedure

→ Method 1 Reacting an acid with an insoluble solid

1 Measure the required volume of acid into a beaker. Add the insoluble metal oxide, metal hydroxide, or carbonate bit by bit until no more dissolves in the acid. Warm if necessary. Adding a slight excess of the solid ensures that all the acid is used up.

solid

stirring
rod

dilute acid

2 Filter off the excess solid, collecting the solution of the salt in an evaporating basin. The **residue** on the filter paper is the excess solid.

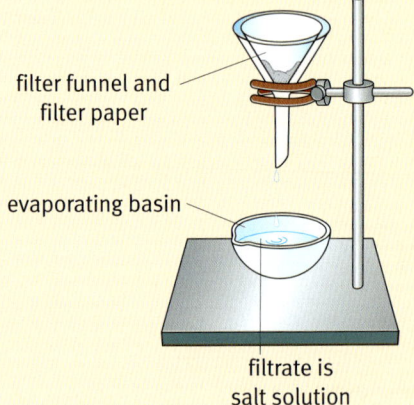

filter funnel and
filter paper

evaporating basin

filtrate is
salt solution

3 Heat gently to evaporate some of the water. Evaporate until crystals form when a droplet of solution picked up on a glass rod cools.

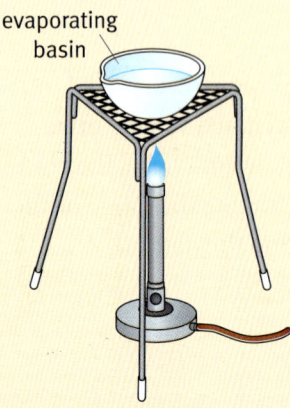

evaporating
basin

4 Pour the concentrated solution into a labelled Petri dish and set it aside to cool slowly and crystallise.

Principle

This procedure is suitable for an alkali or a carbonate that is soluble in water.

Procedure

Method 2 Reacting an acid with an alkali or soluble carbonate

1 Measure the required volume of a solution of the alkali or carbonate into a beaker. Add the acid gradually to the solution in the beaker. Mix well with a stirring rod.

alkali or carbonate

DILUTE ACID

2 Use the stirring rod to take a drop of the mixture and test it with a small piece of indicator paper on a white tile. Continue adding acid until the solution in the beaker is just neutralised.

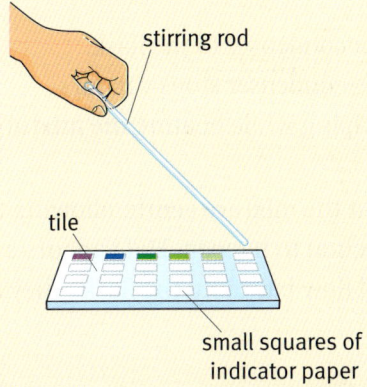

stirring rod

tile

small squares of indicator paper

3 Pour the neutral solution of the salt into an evaporating basin. Heat gently to evaporate. Continue heating until crystals form when a droplet of solution picked up on a glass rod cools.

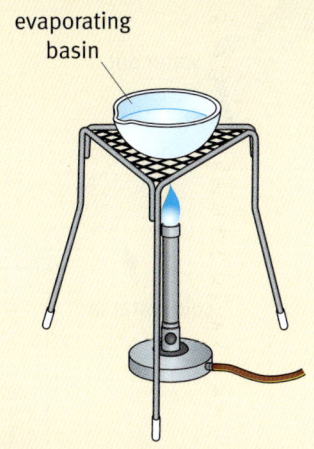

evaporating basin

4 Pour the concentrated solution into a labelled Petri dish and set it aside to cool slowly and crystallise.

Making an ester

⮕ Principle

Alcohols react with **carboxylic acids** to make **esters**.

alcohol + carboxylic acid → ester + water

This type of reaction is very slow at room temperature. Acids such as sulfuric acid catalyse the reaction. Heating also helps to speed up the reaction.

⮕ Procedure

1 Measure the required volumes of alcohol and carboxylic acid into a flask. Then add two drops of sulfuric acid to act as a catalyst.

2 Fit a condenser to the flask. This condenser stops vapours escaping while heating the mixture.

3 Heat the mixture gently, allowing time for the reaction to happen. Hot vapours condense and flow back into the flask. They **reflux**.

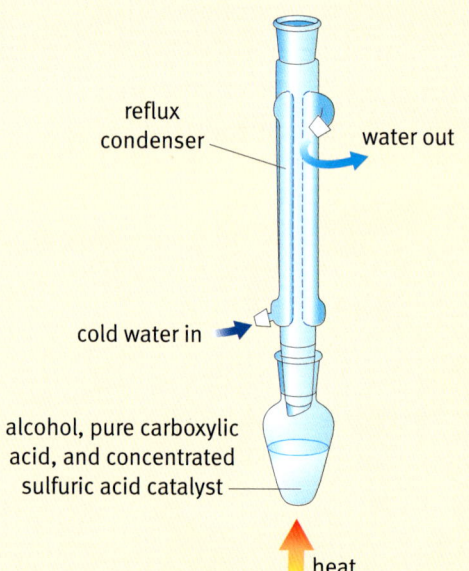

reflux condenser

water out

cold water in

alcohol, pure carboxylic acid, and concentrated sulfuric acid catalyst

heat

4 Rearrange the apparatus to distil off the product. The **distillate** is not pure because it may contain some unchanged alcohol and also some acid.

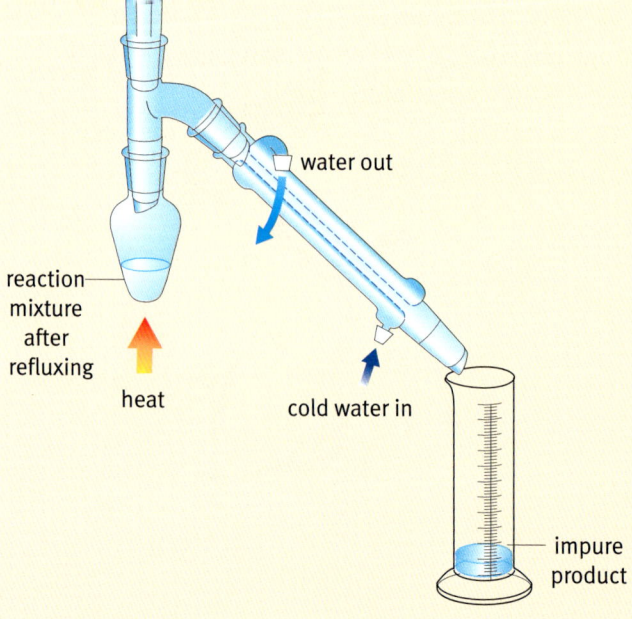

water out

reaction mixture after refluxing

heat

cold water in

impure product

Making an emulsion

CHECK SAFETY
Never work
unsupervised

➡ Principle

An **emulsion** is a mixture of two liquids which do not dissolve in
each other. In an emulsion one of the liquids is very finely dispersed
in the other. Without an **emulsifying agent**, emulsions tend to
separate into two layers. Sometimes the emulsifying agent forms by
chemical reaction during mixing.

➡ Procedure

Procedures for making emulsions vary according
to the ingredients. This is an example of a procedure
to make a cosmetic emulsion.

1 Measure out the oily part of the emulsion.
In this example this is stearic acid.

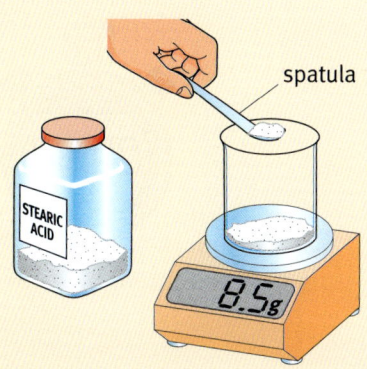

spatula

STEARIC
ACID

8.5g

2 Warm and stir the
stearic acid until it
melts and reaches
the required
temperature.

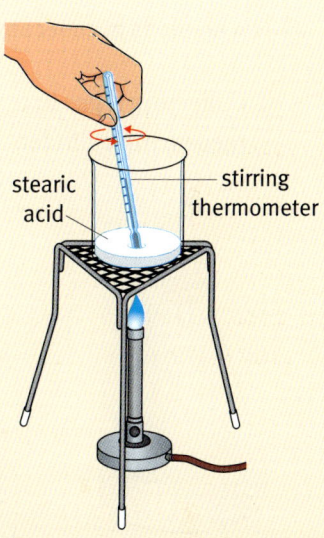

stearic
acid

stirring
thermometer

3 In a second beaker warm
and stir a mixture of
glycerol with a very dilute
solution of alkali.
This makes the watery
part of the emulsion.

glycerol +
a very dilute
solution of
potassium
hydroxide

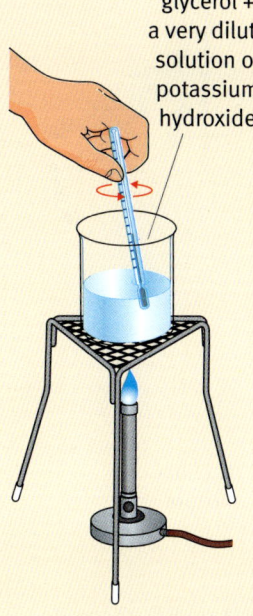

4 Pour the solution of glycerol
in dilute alkali into the
molten stearic acid.
Keep stirring. Some of the
organic acid reacts with
alkali to make the emulsifying
agent. Keep stirring as the
mixture cools and forms a
thick emulsion.

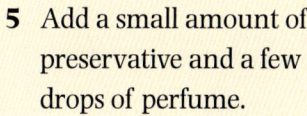

5 Add a small amount of
preservative and a few
drops of perfume.

6 Transfer the emulsion to a
labelled container.

Measuring reaction rates

⇨ Principle

Chemists measure the **rate of a reaction** by finding the quantity of
product produced or the quantity of reactant used up in a fixed time.
The rate can be found by measuring either the rate of disappearance of
one of the reactants or the rate of formation of one of the products.

⇨ Procedures

The diagrams show several methods of measuring
the rate of reaction. The results can be plotted on a
graph to show how the rate varies during the reaction.

⇨ Two methods of collecting and measuring a gas product

1 Collecting a gas in a
measuring cylinder

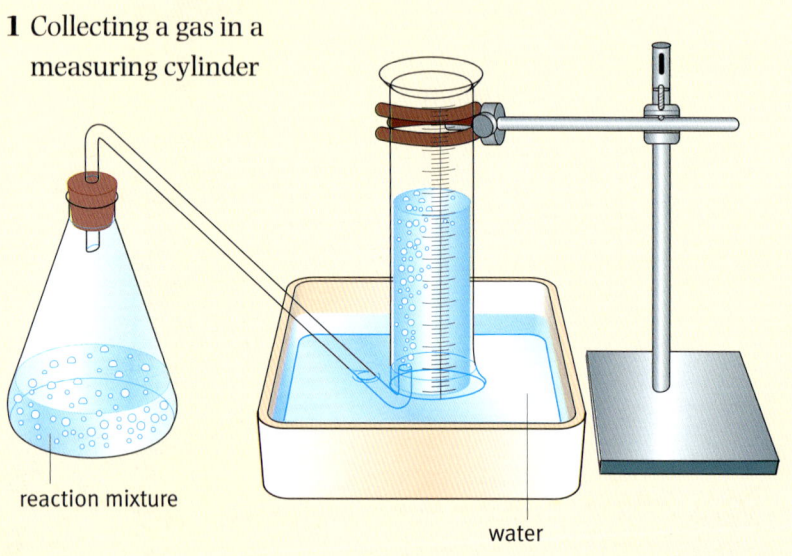

reaction mixture

water

2 Collecting a gas
in a syringe

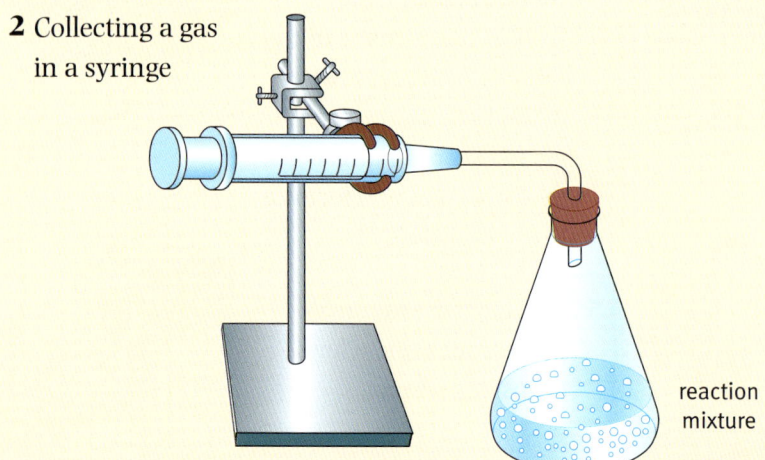

reaction
mixture

Record the volume at regular intervals, such as every 30 or 60 seconds.

➔ Counting the bubble rate as a gas is formed

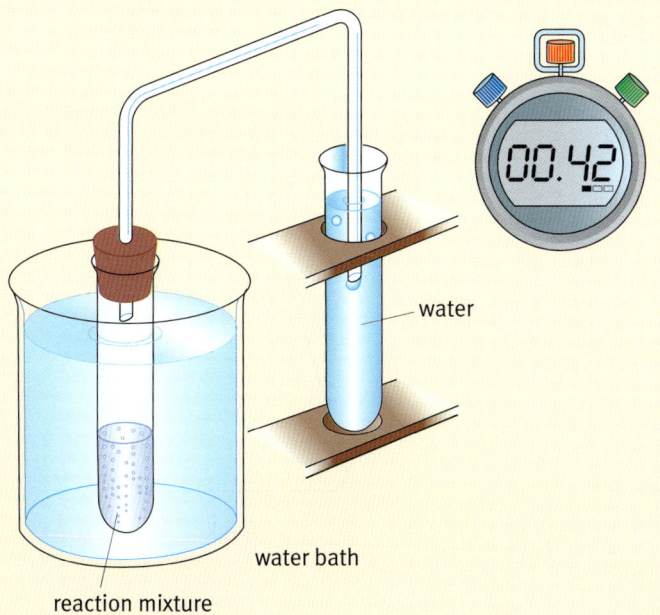

water

water bath

reaction mixture

Count and record how many bubbles are formed during regular time intervals, such as every 20 seconds.

➔ Measuring the loss of mass as a gas is formed

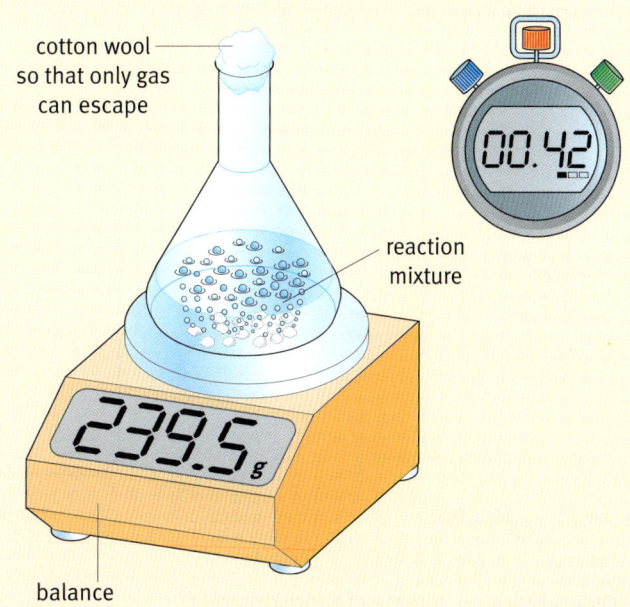

cotton wool so that only gas can escape

reaction mixture

balance

Record the mass at regular intervals, such as every 30 or 60 seconds.

➔ Timing how long it takes for a solution to turn cloudy

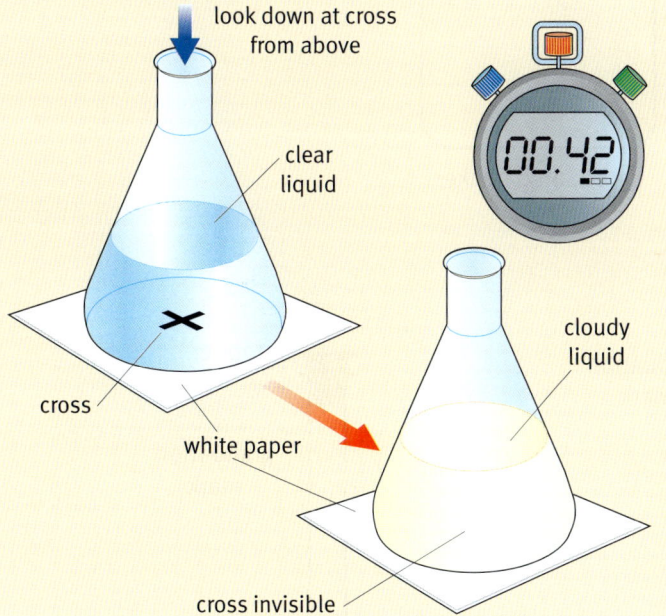

look down at cross from above

clear liquid

cloudy liquid

cross

white paper

cross invisible

This is a method for reactions that produce a precipitate. Mix the liquids in the flask and start the stopwatch. Stop it when you can no longer see the cross.

➔ Timing how long it takes for a solid reactant to dissolve

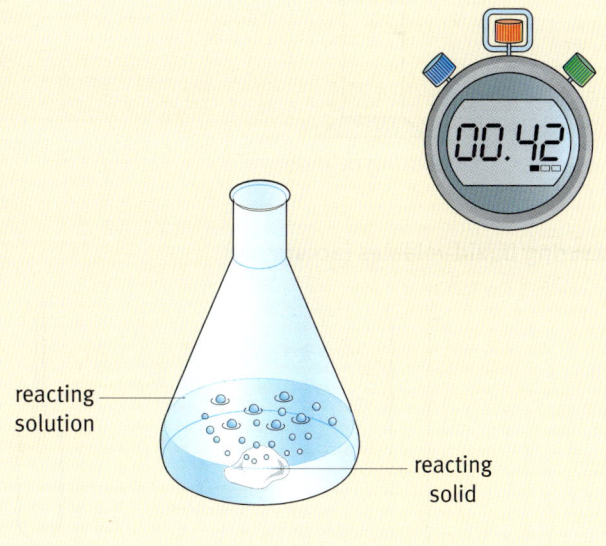

reacting solution

reacting solid

Mix the solid and liquid in the flask and start the stopwatch. Stop it when you can no longer see any solid.

Chemical apparatus

The chemical apparatus on this page and the next are shown in two dimensions. This makes the apparatus much easier to draw.

glass containers

test tube

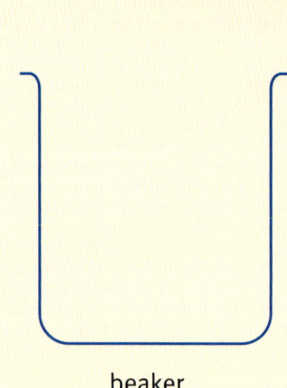

beaker

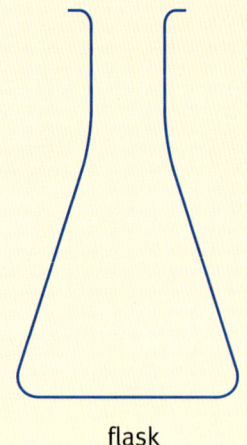

flask

measuring liquid volumes (approximately)

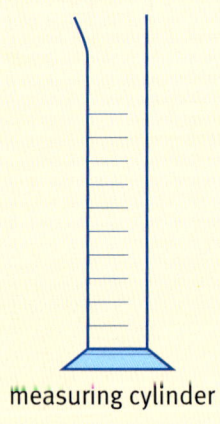

measuring cylinder

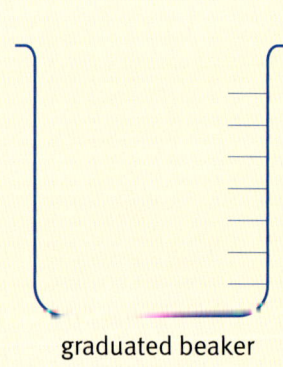

graduated beaker

measuring liquid volumes (accurately)

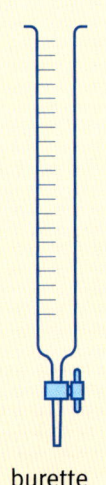

burette

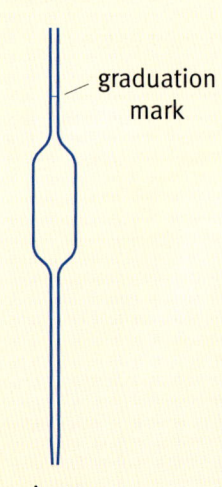

graduation mark

pipette

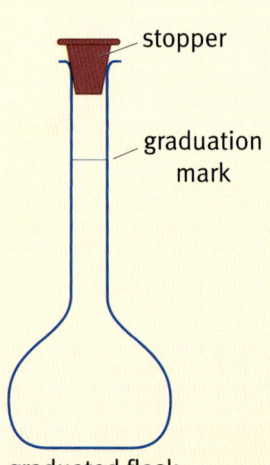

stopper

graduation mark

graduated flask

Heating a liquid

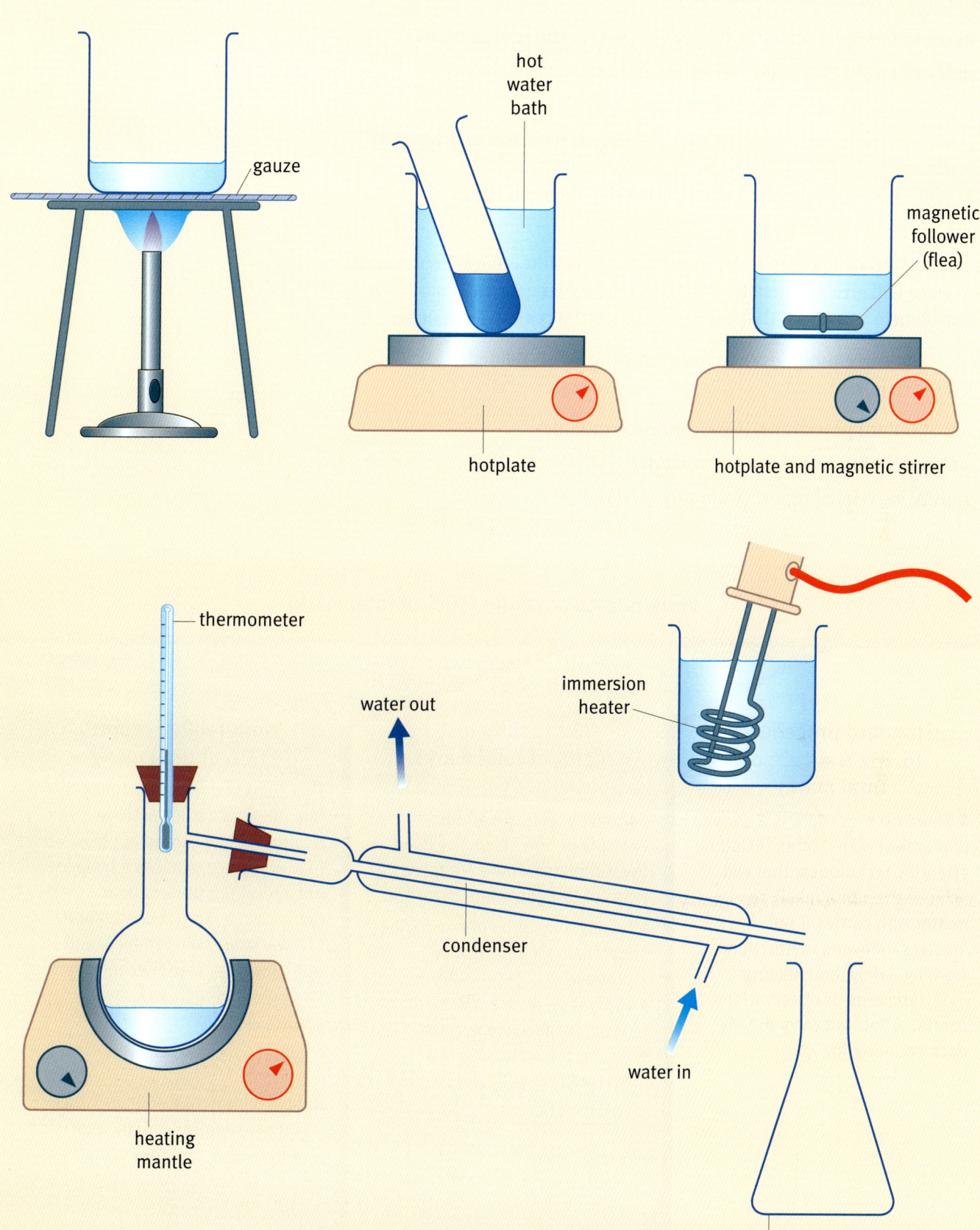

gauze

hot water bath

hotplate

magnetic follower (flea)

hotplate and magnetic stirrer

thermometer

water out

immersion heater

water in

condenser

heating mantle

flask to collect distillate

Your Work-related portfolio

The Additional Applied Science course aims to help you

▶ carry out specific scientific procedures where the results matter

▶ apply science knowledge and techniques, to solve problems

▶ learn about a variety of science-based workplaces

▶ select, organise, and communicate information clearly and logically

You will show your progress with these skills through your Work-related portfolio.

Your Work-related portfolio counts for 50% of the total mark. Your school or college will give you details of the marking scheme for each part of it. This will help you check that your work meets the criteria for success.

Across the three modules that you study, the Work-related portfolio requires

▶ six Standard procedures (two from each module)

▶ one Suitability test (from any module)

▶ one Work-related report (from any module)

Work-related portfolio (50% of total mark)

Standard procedures (6 × 2% = 12% of total mark)

A Standard procedure is a series of practical steps, often including scientific techniques, that will achieve the same result no matter who carries it out. It involves following instructions, working safely, and making measurements or observations carefully. You will carry out six, each counting for 2%.

Suitability test (21% of total mark)

Suitability tests are another example of how science is used in the workplace. There are three types of test you might carry out:

▶ testing a material or comparing materials for a particular purpose

▶ comparing different procedures used for the same purpose

▶ testing the suitability of a device for a particular purpose

Work-related report (17% of total mark)

This gives you the opportunity to find out about some science-related activity carried out in a real workplace, such as a hospital or factory. You present your findings in a Work-related report. Choose your topic carefully, and don't let your report become too large. A good topic will

▶ interest you

▶ contain enough scientific content

▶ have information sources which you can identify, obtain, and understand

Standard procedures

Typically standard procedures will be carried out and assessed in a single lesson. They will involve a series of steps. Your teacher will tell you what, if anything, needs to be recorded.

Suitability test

You will need to

▶ describe desirable properties or characteristics

▶ follow or devise a suitable approach

▶ collect reliable data

▶ evaluate the suitability of the material, procedure, or device

▶ communicate through a structured report

Work-related report

For your Work-related report, you will probably use the following sources of information:

▶ Internet

▶ school library

▶ local public library

▶ TV/video

▶ newspapers and magazines

▶ museums and exhibitions

You may also gather information from specific people or organisations:

▶ interview a practitioner (possibly a family member or a friend)

▶ write a letter to an organisation

▶ telephone to request a leaflet, or to find out who to write to

To obtain useful information from any of these, you will need to prepare detailed questions in advance. Speak (or write) politely and explain who you are and what you are doing.

Use all information sources selectively, picking out parts relevant to your topic and writing in your own words. Keep a detailed record of any information sources you use.

Tip

The best advice is 'plan ahead'. Give your work the time it needs and work steadily and evenly over the time you are given. Your deadlines will come all too quickly, especially as you will have coursework to do in other subjects.

Chemical symbols, formulae, and units

Symbols for chemical elements

Metal element	Symbol	Non-metal element	Symbol
calcium	Ca	carbon	C
magnesium	Mg	chlorine	Cl
potassium	K	hydrogen	H
sodium	Na	nitrogen	N
zinc	Zn	oxygen	O
		sulfur	S

Chemical formulae

Acids

Acid	Formula
hydrochloric acid	HCl
sulfuric acid	H_2SO_4
nitric acid	HNO_3

Antacids

Insoluble (or slightly soluble) antacids	Formula
calcium oxide	CaO
calcium hydroxide	$Ca(OH)_2$
magnesium oxide	MgO
magnesium hydroxide	$Mg(OH)_2$
zinc oxide	ZnO

Soluble antacids (alkalis)	Formula
potassium hydroxide	KOH
sodium hydroxide	NaOH

The 'OH' part of a hydroxide is a pair of atoms that go together. In the hydroxides of calcium and magnesium there are two OHs for every metal atom. So in the formula the OH appears in brackets with a 2 after it.

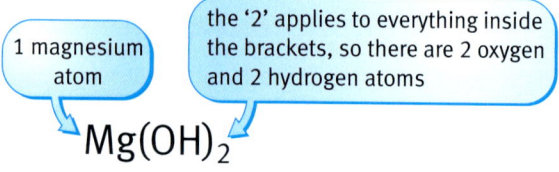

1 magnesium atom

the '2' applies to everything inside the brackets, so there are 2 oxygen and 2 hydrogen atoms

$Mg(OH)_2$

Salts

Parent acid	Related salts	Examples
hydrochloric acid, HCl	chlorides	sodium chloride, NaCl
		potassium chloride, KCl
		calcium chloride, $CaCl_2$
sulfuric acid, H_2SO_4	sulfates	magnesium sulfate, $MgSO_4$
		sodium sulfate, Na_2SO_4
		zinc sulfate, $ZnSO_4$
nitric acid, HNO_3	nitrates	potassium nitrate, KNO_3
		sodium nitrate, $NaNO_3$

Parent metal oxide or hydroxide	Related salts	Examples
sodium hydroxide, NaOH	sodium salts	sodium carbonate, Na_2CO_3
		sodium chloride, NaCl
		sodium sulfate, Na_2SO_4
		sodium nitrate, $NaNO_3$
magnesium oxide, MgO or magnesium hydroxide, $Mg(OH)_2$	magnesium salts	magnesium carbonate, $MgCO_3$
		magnesium sulfate, $MgSO_4$
calcium oxide, CaO or calcium hydroxide, $Ca(OH)_2$	calcium salts	calcium carbonate, $CaCO_3$
		calcium chloride, $CaCl_2$
zinc oxide, ZnO	zinc salts	zinc carbonate, $ZnCO_3$
		zinc sulfate, $ZnSO_4$

Units of concentration

The units for concentrations of solutions are usually one of the following:

▶ grams per litre of solution, g/litre (also written as g/dm^3 or $g\ dm^{-3}$)

▶ grams per millilitre of solution, g/ml (also written as g/cm^3 or $g\ cm^{-3}$)

▶ parts per million, ppm (the same as g/1000 litres)

Note: 1 litre = 1000 ml = $1000\ cm^3$ = $1\ dm^3$

If 10 g of a chemical is dissolved in water to make 500 ml of solution, the concentration is 10 g per 500 ml, which is 20 g per 1000 ml (1 litre), which is 20 g/litre.

Glossary

acid A compound that dissolves in water to give a solution with a pH lower than 7. An acid can be neutralised by a base.

actual yield The mass of product obtained at the end of a reaction.

alcohols A family of organic chemicals that contain the OH functional group.

alkali A compound that dissolves in water to give a solution with a pH higher than 7. An alkali can be neutralised by an acid.

antacid A chemical that neutralises an acid, such as a metal oxide or a metal carbonate. A soluble antacid is an alkali.

aqueous In an aqueous solution, the solvent is water.

balanced symbol equation An equation representing a chemical reaction using symbols and formulae, showing equal numbers of each type of atom each side of the equation.

batch process A process which produces a certain quantity of a chemical in one operation. An industrial batch process is a large-scale version of a synthesis carried out in laboratory glassware. After producing a batch of a chemical, the apparatus is cleaned and used again.

bulk chemicals Chemicals made by the chemical industry on a scale of thousands or millions of tonnes per year, such as ammonia, sulfuric acid, sodium hydroxide, chlorine, and ethene.

carboxylic acids A family of organic chemicals that contain the COOH functional group.

catalyst A chemical that speeds up a chemical reaction without itself being used up in the reaction.

chemical formula A way of describing a chemical that uses symbols for atoms. It gives information about the number of different types of atom in the chemical.

concentration The quantity of a chemical dissolved in a stated volume of solution. For units, see page 59.

continuous process A process for manufacturing chemicals on a large scale in industry which operates for 24 hours a day. Raw materials are constantly fed into the plant and products continuously removed.

corrosive A corrosive substance may destroy living tissues on contact.

crystallisation Producing crystals from a solution by evaporation or cooling.

distillate The liquid that vaporises, condenses again, and is collected during distillation.

distillation Heating a liquid mixture to evaporate one or more components, then condensing the vapours back to liquid. This separates liquids with different boiling points.

drugs Chemicals used in medicine for the treatment, relief, diagnosis, or prevention of disease. People also take drugs for stimulation and relaxation.

element A chemical that cannot be broken down chemically into a simpler one. Each element contains only one type of atom.

emulsifying agent A chemical which is added to an emulsion to prevent the two liquids separating into layers.

emulsion A mixture in which one liquid is dispersed through the other in tiny droplets.

endothermic An endothermic reaction takes in energy.

enzymes Natural catalysts in living organisms, that speed up chemical reactions.

esters A family of organic compounds, formed when organic acids react with alcohols.

evaporate To turn a liquid into its vapour, often without boiling. (Wet things dry by evaporation at room temperature.)

exothermic An exothermic reaction gives out energy.

fine chemicals Chemicals made by the chemical industry in smaller quantities than bulk chemicals, such as drugs and pesticides.

formulation A particular mixture designed to make a product suitable for its use.

fraction Chemicals in crude oil separated from the other fractions by distillation. Each fraction is a mixture of compounds with similar boiling points.

functional group A group of atoms that gives each family of organic chemicals their particular properties. The functional group of alcohols is the OH group, and of carboxylic acids is the COOH group.

hydrocarbons Compounds that contain the elements carbon and hydrogen only.

indicator A chemical that shows whether a solution is acidic or alkaline. For example, litmus turns blue in alkalis and red in acids. Universal indicator has a range of colours which show the pH of a solution.

inorganic chemicals Chemicals made using raw materials obtained from mineral sources (including the sea and air), but not from plants, animals, or their dead remains (such as fossil fuels).

insoluble Describes a chemical that will not dissolve in water (or another named solvent).

irritant An irritant substance is not corrosive but may cause inflammation or lesions to the skin or eyes on contact.

medicine A drug or mixture designed to treat or prevent disease.

mineral A naturally occurring solid material made up mainly of one chemical. Examples of minerals include limestone (calcium carbonate) and rock salt (sodium chloride).

non-aqueous solvent A solvent that does not contain water.

non-renewable resources Resources that cannot be replaced as quickly as they are used, such as fossil fuels

ore A mineral from which a metal can be extracted, e.g. haematite (iron ore) or bauxite (aluminium ore).

organic chemicals Carbon compounds, often originating from living organisms, or from their remains in fossil fuels such as crude oil.

organic chemistry The study of the many different types of carbon compounds.

oxide A compound of an element with oxygen.

percentage yield The percentage of the theoretical yield that was actually obtained in a reaction.

petrochemicals Carbon compounds, often hydrocarbons, that come from fossil fuels such as crude oil. Petrochemicals are used to make plastics, fibres, medicines, dyes, and detergents.

pH scale A number scale which shows the acidity or alkalinity of a solution in water.

pigment An insoluble chemical added to a product such as a paint, cosmetic, or plastic, to give them colour or to make them opaque.

precipitate A solid product which comes out of solution in a chemical reaction.

pure A chemical is pure if it is a single element or compound not mixed with anything else.

rate of reaction A measure of how quickly a reaction is happening. It may be found by measuring the rate of disappearance of a reactant or the rate of appearance of a product. In most chemical reactions the rate changes with time.

raw materials Starting materials used to make a product. The term usually refers to naturally occurring materials, e.g. wood, cotton, coal, petroleum, natural gas, sulfur, limestone, metal ores, and other minerals.

reacting masses The masses of chemicals that react together, and the masses of product that are formed in a reaction. Reacting masses are calculated from the balanced symbol equation using relative atomic masses and relative formula masses.

refinery An industrial plant where crude oil is separated into fractions by distillation, converted into new chemicals, and purified.

reflux A process used to prevent a liquid evaporating away during heating; vapours are condensed and run back into the flask.

relative atomic mass The mass of an atom of an element compared with the mass of an atom of carbon. The relative atomic mass of carbon has been defined as 12. On the same scale the relative atomic mass of hydrogen is 1.

relative formula mass The combined relative atomic masses of all the atoms in a molecule. To find the relative formula mass of a molecule, you just add up the relative atomic masses of its atoms.

renewable resources Resources that can be replaced as quickly as they are used, such as wood.

residue Solid substance that collects on the filter paper during filtration, or is left in the flask after distillation.

salt A chemical formed when an acid reacts with a metal, a metal oxide, or a metal carbonate.

scale up To find a way of producing a chemical on a large scale by adapting the reaction that produced it during research and development.

soluble Describes a chemical that dissolves in water (or another named solvent).

solute Substance (usually a solid) that is dissolved by a solvent to form a solution.

solvent A liquid used to dissolve a substance (the solute).

speciality chemicals Chemicals made by the chemical industry on a small scale, which are needed by other manufacturers for particular purposes. They include flame retardants, food additives, and liquid crystals for flat-screen televisions.

stabiliser A chemical that helps to make permanent food emulsions. Stabilisers make mixtures thicker, which stops dispersed liquid drops coming together.

standard procedure Precise instructions written so that scientists carry out a preparation, analysis, or test in the same way every time.

suspension A mixture of an insoluble solid with a liquid. The solid particles are dispersed through the liquid. You cannot see through a suspension.

sustainable A sustainable process, like sustainable manufacture, can carry on without harming the environment over a long period of time.

synthetic routes Methods of synthesising a chemical using different series of chemical reactions or different starting reactants.

theoretical yield The amount of product that would be obtained in a reaction if all the reactants were converted to products exactly as described by the balanced chemical equation.

word equation A summary in words of a chemical reaction.

yield The amount of product obtained from a chemical reaction. It may be measured as the actual yield or the percentage yield.

Index

UNIVERSITY PRESS

Great Clarendon Street, Oxford OX2 6DP

Oxford University Press is a department of the University of Oxford.
It furthers the University's objective of excellence in research, scholarship,
and education by publishing worldwide in

Oxford   New York

Auckland   Cape Town   Dar es Salaam   Hong Kong   Karachi
Kuala Lumpur   Madrid   Melbourne   Mexico City   Nairobi
New Delhi   Shanghai   Taipei   Toronto

With offices in

Argentina   Austria   Brazil   Chile   Czech Republic   France   Greece
Guatemala   Hungary   Italy   Japan   Poland   Portugal   Singapore
South Korea   Switzerland   Thailand   Turkey   Ukraine   Vietnam

© University of York on behalf of UYSEG and the Nuffield Foundation 2006

British Library Cataloguing in Publication Data

Data available

ISBN-13: 978-0-19-915029-8
ISBN-10: 0-19-915029-X

10 9 8 7 6 5 4 3 2

Design by IFA Design, Plymouth, UK

Printed in Italy by Rotolito Lombarda

THE
WELLCOME
TRUST

The
Nuffield
Foundation

Acknowledgements

We are very grateful to the teachers and students in pilot schools for their detailed and constructive
recommendations for the revisions to this module.

We would also like to thank Mary Whitehouse and the examining team that developed the specification
for this module.

The authors drew on the invaluable work of the Chemical Industry Education Centre at York
(http://www.uyseg.org/ciec_home.htm) when developing the resources for this module.

We thank Barry Taylor of Danisco Foods for his help with a case study in this book.

The publisher would like to thank the following for their kind permission to reproduce copyright material:

p2 Simon Fraser/Pharmacy Department, Rvi, Newcastle Upon-tyne/Science Photo Library; **p3** Chris
Hellier/Corbis UK Ltd.; **p4** Roy Morsch/Corbis UK Ltd.; **p5** Heather Angel/Natural Visions; **p5** Th Foto-
werbung/Science Photo Library; **p6** Ghislain & Marie David de Lossy/The Image Bank/Getty Images; **p6** Zooid
Pictures; **p7** R. Maisonneuve/Publiphoto Diffusion/Science Photo Library; **p8** Maureen Barrymore/Corbis UK
Ltd.; **p9** Lawson Wood/Corbis UK Ltd.; **p10l** Robert Brook/Science Photo Library; **p10tr** TH Foto/Alamy;
p10cr Arnold Fisher/Science Photo Library; **p10br** Tony Waltham/Geophotos; **p11t** R. Estall/Robert Harding
Picture Library Ltd/Alamy; **p11b** Les Polders/Alamy; **p12** Maximilian Stock Ltd/Science Photo Library;
p13l Geoff Tompkinson/Science Photo Library; **p13r** William Taufic/Corbis UK Ltd.; **p14t** Esso Petroleum
Company Limited/ExxonMobil; **p14b** Esso Petroleum Company Limited/ExxonMobil; **p15l** Huntsman Tioxide;
p15r Leslie Garland/Leslie Garland Picture Library/Alamy; **p16** Zooid Pictures; **p17** Colin Cuthbert/Science
Photo Library; **p18** Robert Brook/Science Photo Library; **p19** Tek Image/Science Photo Library; **p20** Danisco;
p21 Danisco; **p22** John Kaprielian/Science Photo Library; **p24l** Martyn F. Chillmaid; **p24c** Zooid Pictures;
p24r Martyn F. Chillmaid; **p25** Andrew Lambert Photography/Science Photo Library; **p26** Zooid Pictures;
p27 Martyn F. Chillmaid; **p29tl** Vincent van Gogh Foundation/Van Gogh Museum Enterprises B.V.;
p29cl Richard Megna/Fundamental Photos/Science Photo Library; **p29bl** Zooid Pictures; **p29tr** Medical-on-
Line; **p29cr** Zooid Pictures; **p29br** Zephyr/Science Photo Library; **p31** JeffMcIntosh/CP/AP/Empics; **p32** Andrew
Syred/Science Photo Library; **p33** Zooid Pictures; **p37** Martyn F. Chillmaid; **p38** Crown Copyright Health &
Safety Laboratory /Science Photo Library; **p40** Gary Banks/BP Saltend; **p41** Danny Lehman/Corbis UK Ltd.;
p44l Zooid Pictures; **p44c** Zooid Pictures; **p44r** Zooid Pictures; **p44** Ron Boardman/Frank Lane Picture
Agency/Corbis UK Ltd.; **p45l** Martyn F. Chillmaid; **p45r** Martyn F. Chillmaid; **cover** OUP/Photodisc

Illustrations by IFA Design, Plymouth, UK and Clive Goodyer